Introduction to the Maths and Physics of the Solar System

Introduction to the Maths and Physics of the Solar System

Lucio Piccirillo

CRC Press
Taylor & Francis Group
Boca Raton   London   New York

CRC Press is an imprint of the
Taylor & Francis Group, an **informa** business

First edition published 2020
by CRC Press
6000 Broken Sound Parkway NW, Suite 300, Boca Raton, FL 33487-2742

and by CRC Press
2 Park Square, Milton Park, Abingdon, Oxon, OX14 4RN

© 2020 Taylor & Francis Group, LLC

CRC Press is an imprint of Taylor & Francis Group, LLC

International Standard Book Number: 978-0-367-02271-6 (Hardback)
International Standard Book Number: 978-0-367-00254-1 (Paperback)

CV 03.18.2020 1320

To Emma, Julia, Madelaine, Jacopo and Tommaso.

This book is also dedicated to Mario Sortino, a.k.a. "Zio Mario", for explaining to me the foundation of the scientific methods during long and extenuating mountain hikes during my teenager years.

Finally, I would like to express my deepest affection and acknowledgement to a very special person, CB, for her lovely support in the final phases (the most stressful) of the writing. Thank you CB!

Contents

Foreword

This book collects my experiences after teaching an introductory undergraduate course on the physics of the solar system at the University of Manchester. The course is oriented to first-year undergraduate students first semester, i.e. students just attending a university course for the first time. My philosophy of teaching has been always to teach trying to achieve several objectives.

First, the course has to trigger interest in the student: this is relatively easy because space, planets, etc., are a natural interesting subject for practically everybody. Who does not feel a sense of awe when thinking about the Universe?

Second, students have the opportunity to learn how natural phenomena are first observed carefully then modeled and finally calculated and estimated more than just simply described. Every time there is a discovery in astrophysics, newspapers are full of sensational reports of phenomena described almost as mysteries of nature. Scientists are seen as "magicians" that mysteriously know what a black hole is or what space-time is. First year students are usually still in this mode of "acceptance" of qualitative explanations more than critical evaluation and quantitative estimation.

Third, I wanted to write a book that is somehow "self-contained", i.e. all the maths and physics contained are explained starting from basic concepts that any well-educated person can follow. I imagine my reader as a person who has the curiosity to understand quantitatively how scientists can make claims that apparently are "magic" like, for example, stating that the age of the Earth is about 4.5 billion years. In this book, this person, with a little bit of effort, will be able to convince himself or herself that the claim is perfectly justified.

Lastly, when I was a young high school student I always wanted to read about science but I also wanted to be "convinced" about the various scientific claims. I never found a book that was easy enough to describe all the maths and physics needed but complete enough to describe phenomena accurately. About 45 years ago I promised to myself that, if I was capable, I would have written such a book and here it is. Now I need a time machine to send it to myself...

Preface

This book is an attempt at having a self-contained treatment of some phenomena happening in the solar system. Self-contained because all the maths and physics needed are described starting from a few basic notions that every logical person should have no difficulty to accept and digest. Instead of a mere qualitative description of various phenomena, I tried to select a set of representative phenomena happening in our solar system and describe them quantitatively after a careful treatment of the maths and physics needed to understand.

A correct title of this book should be: "How to Teach You Maths and Physics Concepts Using Interesting Facts about the Solar System". The first chapter, for example, is designed to introduce all the geometry, trigonometry, calculus, and vector calculus needed to understand planetary orbits and more. I start the chapter by studying in detail Eratosthenes determination of the circumference of the Earth and I show how such a simple geometry problem contains in it a big number of assumptions. When determining an important angle through inverting the sine function instead of just stating that an angle α is such that its sine is 0.1256, I use this opportunity to introduce derivatives and series expansion of functions. The maths is obviously treated without the rigor usually associated with more complete and exhaustive textbooks that the reader is warmly invited to consult.

My experience is that learning maths and physics using interesting problems gives the students, and I hope more generally the readers of this book, additional enjoyment by showing how a variety of phenomena can be, and are, calculated when the theory that we have is a good theory.

This approach has the risk that the book might be "too complex" for people who do not have a technically oriented mind and "too easy" for people that already know the basics. My writing is therefore always trying to excite anyone who already "knows" and allowing to understand who does not know "yet". Who already "knows" might find, here and there, interesting new calculations and perhaps different views of already known facts. Those who, instead do not know enough, might still get the opportunity to try to understand how a physicist works: I only ask some efforts in following carefully my maths and physics explanations.

I strongly believe that science is fun, simple maths used in physics is understandable, and everybody equipped with patience, enthusiasm, and time can enjoy the beauty of using maths to "predict" phenomena happening in the solar system.

This book is my private acknowledgment of my hero, Galileo Galilei. He was the first to understand that "the book of nature is written in the language of mathematics".

Basic Concepts

CONTENTS

T HERE are several elementary concepts that need to be properly reviewed in order to fully understand the phenomena happening in our
solar system. Obviously, we do not have here the space or the time to review
the entirety of geometry, trigonometry, calculus, etc. However, especially for
those readers that need to refresh concepts that haven't been used in some
time, we treat in some detail most of the useful mathematical concepts that
will be used in later chapters, especially the chapter on celestial mechanics.
Our treatment is by no means rigorous and is mostly used to prove statements
about our world and, more specifically, our solar system. We will see in the
next few sections how the knowledge of very simple geometry has allowed
ancient Greek philosophers to state that not only the Earth is spherical, but
also estimate the distance from the Earth to the Moon and the Earth to the
Sun.

We will use the Greek philosophers' amazing achievements as an excuse
to introduce some of the concepts that are needed in the rest of the book
putting considerable attention to show how maths is deeply used – we would
say *embedded* – in the physical world. We will also often digress to show
how systems and concepts apparently disconnected from the study of the
solar system are actually used, sometimes so automatically that we forget
how important they are. An example is the case of digital electronics and
logic circuits inside computers.

1.1   GEOMETRY

Now imagine you are a Greek philosopher, named Eratosthenes (see fig. 1.1), residing in a beautiful city in Egypt (Alexandria). You are taking care of one of the most complete and important libraries in existence and you have time to read, study and think. Somebody has just told you that during one of his trips to Syene (today's Aswan) in the south of Egypt, he has noticed something weird: at midday of the special day of summer solstice, i.e. the longest day and shortest night of the year, the Sun shone directly down a deep vertical well. In other words, looking directly down the well, your head blocks exactly the reflection of the Sun by the water at the bottom of the well. The majority of people when confronted with the news would simply think "that's strange" and then go back to their lives. But you have time to think ... and you run outside at midday of the summer solstice in your city of Alexandria and plant a vertical stick in the ground to check if the Sun is overhead in Alexandria as well. It is not! You have time to think ... and all of a sudden you know what all means: the Earth is not flat, it is a sphere! Not only that, but you are capable of calculating the circumference of the Earth, as well. You just need a bit of geometry.

Geometry is a beautiful construction of the human mind. The etymology comes from the Greek word $\gamma\epsilon\omega\mu\epsilon\tau\rho\iota\alpha$ (*geo* - earth, *metron* - measurement). Geometry is the study of the properties and relations among a set of elementary concepts like points, lines, surfaces, solids. We can construct many different geometries depending on the basic assumptions that we make. These basic assumptions are called *postulates* and are given without proof. The most familiar kind of geometry is *Euclidean Geometry* which is based on Euclid's postulates [6] given below:

1. A straight line segment can be drawn joining any two points.

2. Any straight line segment can be extended indefinitely in a straight line.

3. Given any straight line segment, a circle can be drawn having the segment as radius and one endpoint as center.

4. All right angles are congruent[1].

5. If two lines are drawn which intersect a third in such a way that the sum of the inner angles on one side is less than two right angles then the two lines inevitably must intersect each other on that side if extended far enough. This postulate is equivalent to what is known as the parallel postulate.

The fifth postulate disturbed Euclid at the point that, wherever possible, he used only the first four. The usage of the fifth postulate gives rise to the so-called *Euclidean Geometry*, i.e. the intuitive geometry that we have of space

---

[1]Congruent means exactly equal in size and shape.

eratosthenes

FIGURE 1.1 Eratosthenes (276 BC–194 BC) was a Greek mathematician who enjoyed writing about music, astronomy and poetry. He was the first to calculate the circumference of the Earth based on simple observations (see text).

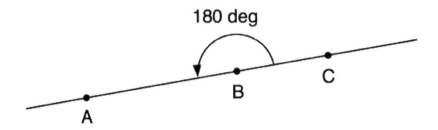

FIGURE 1.2 Straight Angle Theorem.

(also-called *flat* or *parabolic* geometry). If we do not accept Euclid's fifth postulate we can construct other geometries that are completely self-consistent. By stating a different fifth postulate, Lobachevsky and Bolyai-Gauss constructed the so-called *hyperbolic geometry* while Riemann constructed an elliptic geometry. We will need a different geometry when we observe the sky because the stars appear to be fixed to a gigantic rotating sphere. This means that we need to consider the elementary concepts of points and lines now expressed on the curved surface of a sphere. We will review this special geometry later on in the book.

Theorems are statements that are proved to be true and are derived from the postulates using logical steps. As an example of a theorem, let's enunciate the Straight Angle Theorem .

Straight Angle Theorem: *Given three distinct points A, B and C, then B lies between A and C if and only if* $\angle ABC = 180°$.

This theorem states something that appears obvious after looking at fig. 1.2 and we refer the reader to any good geometry book for the proof. Although apparently obvious, the Straight Angle Theorem plays an important role in most subsequent proofs. The next theorem – the Interior Angle Theorem – was also used by Greek philosophers to estimate the radius of the Earth. Its proof makes use of the Straight Angle Theorem (1.3).

Interior Angle Theorem: *If two parallel lines are cut by a transverse, then the pairs of alternate interior angles are congruent.*

Let us discuss one of the most famous theorems attributed to the Greek mathematician Pythagoras.

Pythagorean Theorem: *The square of the hypotenuse of a right triangle is equal to the sum of the squares of the other two sides.*

The Pythagorean Theorem is perhaps one of the first ever geometric theorems that we all have studied at school. There are many ways to show its validity and we choose here to show an algebraic-geometric proof.

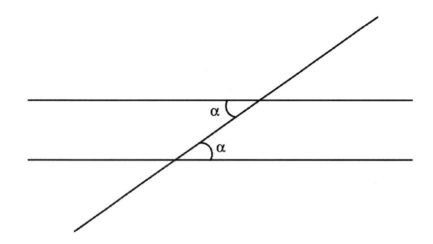

FIGURE 1.3 Interior Angle Theorem: two parallel lines are cut by a transversal. The two interior angles α are congruent.

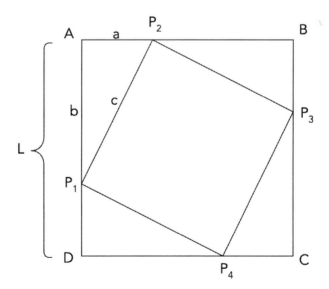

FIGURE 1.4 Proof of Pythagorean theorem.

We start by looking at fig. 1.4. Start by drawing a square and identify a point on one of its sides such that $L = a + b$. This square will have an area equal to $L^2 = (a+b)^2$. By construction, the area of each of the 4 right triangles is $\frac{1}{2}ab$. We can now express the area of the large square in two different ways. The first would be:

$$A = L^2 = (a + b)^2 = a^2 + B^2 + 2ab \tag{1.1}$$

The second way would be by adding the area of the square of side c to the areas of the four right triangles of sides a, b and c.

$$A = c^2 + 4\frac{1}{2}ab = c^2 + 2ab \tag{1.2}$$

equating equations 1.1 and 1.2 we have:

$$a^2 + b^2 + 2ab = c^2 + 2ab \tag{1.3}$$

after eliminating the two common terms in both sides of eq. 1.3, we are left with:

$$c^2 = a^2 + b^2 \tag{1.4}$$

for any right triangles.

Let us see how Eratosthenes of Cyrene estimated the circumference of the Earth about 200 BC (see fig. 1.11). Take into account that Eratosthenes probably used different enunciation of the theorems. We will use here the modern symbols and theorems.

There is an important observation made by Eratosthenes that allowed him to make his extraordinary claim. At midday in Syene the Sun was exactly overhead. At exactly the same midday in Alexandria the Sun is **not** exactly overhead. In fact, he measured the length of the shadow cast by a vertical stick at the same time as the Sun was perfectly vertical in Syene. He noticed that if he can measure the length of the shadow, knowing the height of the stick he can calculate the angle α in figs. 1.5 and 1.11. In order to do so he had to solve a triangle where he knew two sides (the length of the stick and the length of the shadow) and a non-included angle (see the small triangle at the bottom right of fig. 1.5). Notice that we can assume that the shadow is perpendicular to the light rays coming from the Sun.

To calculate the angle α we need some trigonometry.

1.2 TRIGONOMETRY

To find the angle α we need to review the concept of sine and cosine of an angle. We refer to fig. 1.7 representing a circumference of unit radius $\overline{OP} = 1$. The center of the circumference is also the origin of a Cartesian coordinates system (see fig. 1.6) with axes x and y. Let's consider the point P of intersection between the circumference and the radius. We can project the point P

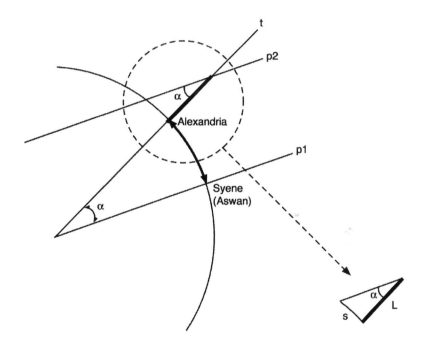

FIGURE 1.5 Eratosthenes determination of the Earth's circumference. Determining the angle α.

FIGURE 1.6 René Descartes (1596-1650) was a French mathematician and philosopher. He was the first to develop the field of analytical geometry and introduced the so-called Cartesian plane.

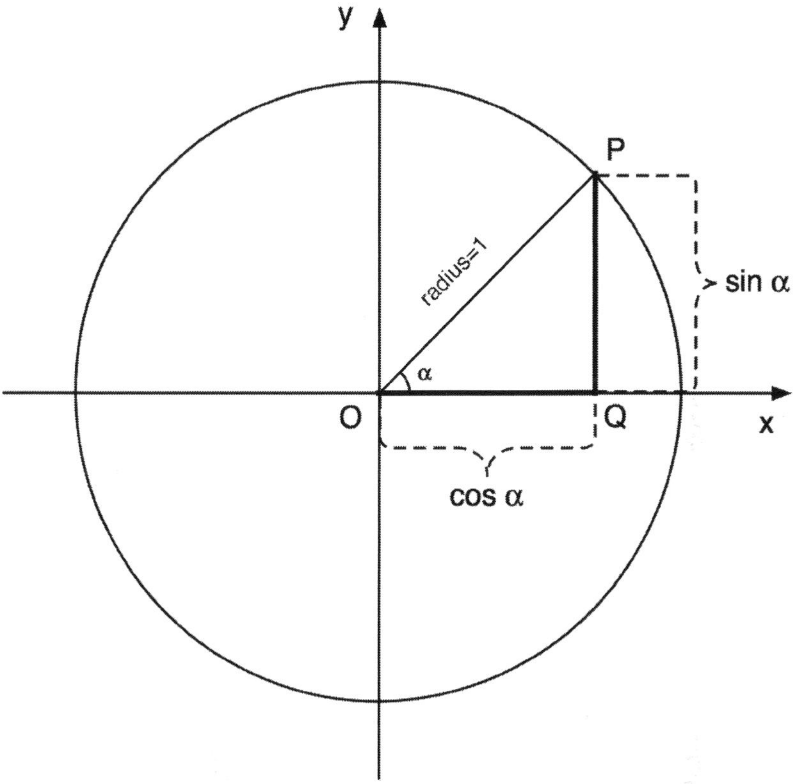

FIGURE 1.7 Sine and cosine defined on the unit circle.

perpendicular to the x axis to the point Q. Because of the projection the angle $\angle OQP$ is equal to 90°. Notice that the lengths of the segments $\overline{OQ}$ and $\overline{QP}$ depend on the angle α. In particular, when α is equal to zero then $\overline{OQ} = 1$ and $\overline{PQ} = 0$. We define the function *sine* of the angle α ($\sin \alpha$), the length of the segment $\overline{PQ}$, and the function *cosine* of the angle α ($\cos \alpha$) the length of the segment $\overline{OQ}$. It follows immediately that for any angle α, we have:

$$sin^2\alpha + cos^2\alpha = 1 \qquad (1.5)$$

Equation 1.5 expresses the Pythagorean Theorem (see eq. 1.4) on the triangle OPQ (see fig. 1.8). It is easy to see that the sine of $\alpha = 0$ is equal to 0 while the sine of $\alpha = 90°$ is equal to 1.

Can we define the functions sine and cosine if the circle is not of radius equal to 1? Obviously yes. With reference to fig. 1.7, suppose that the radius of the circle is not unitary, then in general we have the following definitions:

FIGURE 1.8 Pythagoras (570-495 BC) was one of the most important Greek philosophers. He set the basis of Western Philosophy. He is thought to have proposed many important theorems in geometry the most famous of which is his Pythagorean Theorem on square triangles.

$$\sin \alpha = \frac{\overline{PQ}}{\overline{OP}} \tag{1.6}$$

$$\cos \alpha = \frac{\overline{OQ}}{\overline{OP}} \tag{1.7}$$

Therefore, when the radius is not unity, to correctly express the sine of the angle, we have to divide by the length of the radius.

Therefore, in general, in a right triangle the sine of an angle α is obtained by taking the ratio of the side **opposite** to the angle α to the hypotenuse. Equally, the cosine of an angle α is obtained by taking the ratio of the side **adjacent** to the angle α to the hypotenuse. It follows that sine and cosine functions are strictly related. If we consider a right triangle, we have seen above that if the hypotenuse has a length equal to 1, then the sine and the cosine of an angle are interpreted respectively as the projection to the y and x axes of a Cartesian coordinate system. Now let us study fig. 1.9. Let's project the point P to the y axis on the point Q'. As a result of the projection, the line $\overline{PQ}$ is parallel and equal to the line $\overline{OQ'}$, while the line $\overline{PQ'}$ is equal and parallel to the line $\overline{OQ}$.

We now show that the following two relationships are true:

$$\cos (90 - \alpha) = \sin \alpha \tag{1.8}$$

$$\sin (90 - \alpha) = \cos \alpha \tag{1.9}$$

In order to do so, see fig. 1.9. The segment $\overline{PQ} = \overline{OQ'}$ is opposite to the angle α and therefore is the $\sin \alpha$. The segment $\overline{OQ} = \overline{PQ'}$ is adjacent to the angle α and therefore is the $\cos \alpha$. The angle $\angle Q'OP$ is equal to $90° - \alpha$ because the angle $\angle Q'OQ$ is a right angle. We observe now that the segment $\overline{OQ'}$ is adjacent to the angle $90° - \alpha$ and therefore it is the $\cos(90° - \alpha)$. At the same time, the same segment $\overline{OQ'}$ is opposite to the angle α which means that it is equal to the $\sin \alpha$. We have just proved the first of the eq. 1.9. The other equation can be proved by doing a similar reasoning on the segment $\overline{Q'P}$.

We now give another very useful relationship: the law of sines .

Law of sines: Given the sides A, B and C and angles a, b and c of a triangle (see fig. 1.10), the law of sines is the following equation 1.10:

$$\frac{A}{\sin a} = \frac{B}{\sin b} = \frac{C}{\sin c} \tag{1.10}$$

With reference to fig. 1.5, we call the height of the stick L, the length of its shadow s. Assuming that the shadow is perpendicular to the light rays, it means that the angle opposed to the tower is a right angle. Now let us find the angle α using the law of sines.

In fig. 1.5 the triangle made of the stick, its shadow and the non-included

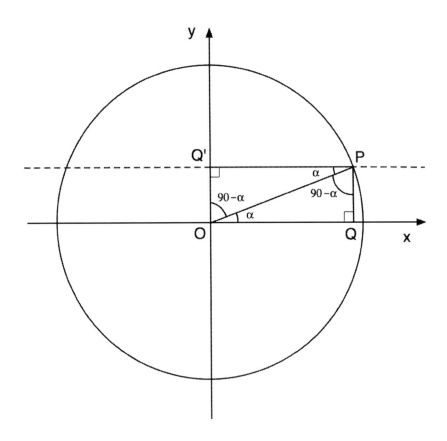

FIGURE 1.9 Sine and cosine relationships.

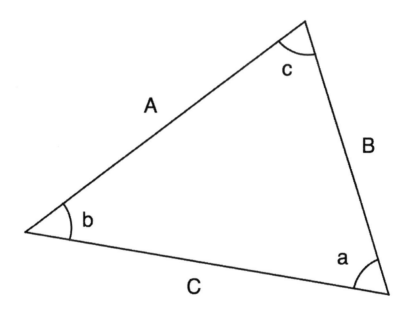

FIGURE 1.10 Generic triangle with sides A,B,C and angles a,b, and c.

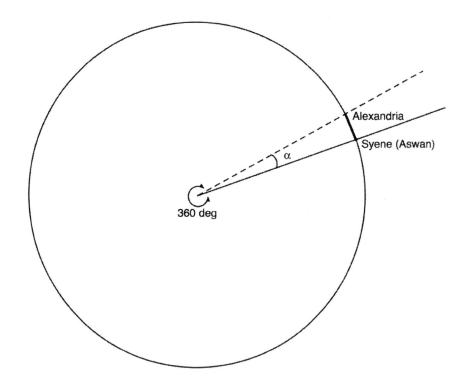

FIGURE 1.11 Geometry used by Eratosthenes. Knowing the angle α and the distance between Alexandria and Syene, the circumference of the Earth can be estimated (see text).

angle α is evidenced. With a good approximation we can assume that the shadow s is perpendicular to the stick L, i.e. the angle $\angle sL$ is equal to $90°$. We now know how to calculate the sine of the angle α as the ratio of the hypotenuse (found using the Pythagorean Theorem) to the length of the shadow s. Eratosthenes found the ratio to be $\sin\alpha = 0.1256$. It is important to underline that the angle α in the function sine is expressed in radians and not degrees!

When an angle is obtained from a ratio of two segments, it is expressed in *radians*. One radian is the angle subtended at the center of a circle by an arc that is equal in length to the radius of the circle. It follows that an angle of $360°$ is equal to 2π rad where $\pi = 3.1415927....$

So, we have the sine of an angle. How do we obtain the angle if we know its sine? Or, in other words, what is the inverse function of the sine? A good scientific calculator has the *inverse sine* function built in. But in order to understand it, we need calculus and Eratosthenes did not have a pocket calculator after all. He had to do all his calculations by hand. We are lucky to live in an era where a computer can do calculation at a speed and volume not accessible to any human beings.

1.3 CALCULUS

In the previous section we followed the logical development of the calculation of the Earth's circumference and we found that we need to evaluate the inverse of trigonometric functions. Let us ask: "how does a pocket scientific calculator evaluate trigonometric and inverse trigonometric functions?" Modern pocket calculators do much more, but for now, let's see how the calculation is performed. The trick is to express complex functions as the sum of many (often infinite) simple functions that can be evaluated easily. Assuming that we can build electronic circuitry that can add and subtract and if we can express any function as a series of additions and subtractions, then our pocket computer can evaluate any function, including the inverse trigonometric functions we are after. We need to be sure that **any** trigonometric function, including its inverse, can be expressed as a series of simple functions that can be evaluated by just adding and subtracting numbers. Let's first see briefly how simple electronics can add and subtract. Once we are convinced that it is possible, we will study how to express functions as series of additions and multiplications.

It turns out that electronic circuits can easily perform additions and subtractions. The arithmetic that we learn very early at school is based on 10 digits, from 0 to 9, probably because we have ten fingers. Electronic circuits work by modifying the flow of an output current (mostly electrons) depending on some input conditions. In other words, for our simple treatment here, an electronic circuit can be schematized as a black box with one or more inputs and one or more outputs. The outputs depend causally on the status of the various inputs. We can try to implement arithmetic in electronic circuits by associating a different voltage (for example) to each of the 10 digits. For ex-

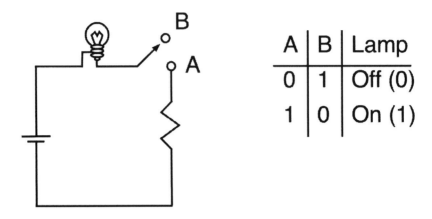

A	B	Lamp
0	1	Off (0)
1	0	On (1)

FIGURE 1.12 Electronic representation of the binary digits "1" and "0". Current flows and a lamp is lit when the switch is in the position "A" or binary condition "1". The lamp is off when the switch is in position "B" corresponding to the binary digit "0".

ample, we could associate 0 Volts to the digit "0", 1 Volt to the digit "1", and so on up to 9 Volts to the digit "9". So a simple 2 digit adder then would be a sort of black box circuit that has two inputs and one output such that the output voltage is the *sum* of the input voltages. The first scientists trying to build a computing machine based on electronic circuits, found out that it is much easier if we change our numbers from base-10 to base-2, i.e. a base where we have only two digits "0" and "1". This arithmetic is called *binary*. It turns out that it is much easier and efficient to build electronics that consider only two voltages, for example 0 Volts and 5 Volts, corresponding to the digits "0" and "1". A binary number is then a long and tedious collection of 0s and 1s like, for example, the number 110101 which is the decimal 53.[2]

Simple electronics, called *gates* can do simple operations like addition of two binary numbers. A computer or a pocket calculator takes your decimal numbers, converts them into binary, does the arithmetic operation in binary form, takes the result in binary and transforms it back to decimal, and finally spits it out on your screen.

---

[2] A decimal number is expressed as a sum of increasing powers of 10. The number 53, for example, is expressed as $5 \cdot 10^1 + 3 \cdot 10^0$. We can express numbers using other bases instead of the number 10. In general we can express any number with a polynomial on base b as $a_0 b^0 + a_1 b^1 + a_2 b^2 + \dots$. If $b = 2$ then we are expressing numbers in base 2. For example the number 110101 is equal to $1 \cdot 2^5 + 1 \cdot 2^4 + 0 \cdot 2^3 + 1 \cdot 2^2 + 0 \cdot 2^1 + 1 \cdot 2^0 = 32 + 16 + 4 + 1 = 53$.

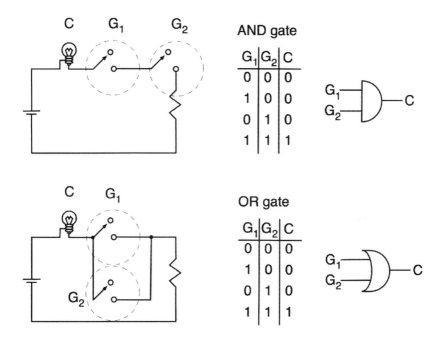

FIGURE 1.13 Realization of the logical gates AND and OR through the usage of switches. The logical symbols are also shown.

Let's explore the basics of digital electronics because we believe that any scientist using a computer should know what is happening inside his/her machine, or at least have a basic understanding. The fundamental unit in digital electronics is the so-called *gate*. A simple electronic circuit that represents binary digits is shown in fig. 1.12. A binary digit "1" can be represented by the lamp "on" while the binary digit "0" by the lamp "off". In fig. 1.12 the switch position "A" allows current to go through the resistor and the lamp and is associated with the digit "1". The other switch position "B" interrupts the current causing the lamp to switch off and is therefore associated with the digit "0".

Our objective is to show how to build a simple circuit that is capable of adding binary numbers. We will use combinations of switches to show how basic gates work and then we will show the final schematic of a simple full adder. In order to do so we need to introduce a few concepts. In fig. 1.13 we see the realization of logical gates. The top circuit is called the AND gate while the bottom gate is called OR. The top circuit shows that the lamp will be switched on if and only if **both** switches are closed or, if switch G_1 AND switch G_2 are closed. If we identify the output C as the status of the lamp,

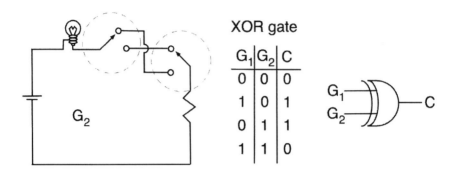

FIGURE 1.14 Exclusive OR (XOR) gate. Only when either G_1 OR G_2, but not both, are "1", then the output is "1". Remember that "1" is when the switch is UP, therefore both switches UP or DOWN do not allow current to flow.

"0" for off and "1" for on, then the switch circuit of the AND gate will obey the AND gate table shown. The AND table tells us that in order to switch on the lamp, both switches must be closed or, symbolically, to have a "1" at the output C, both inputs G_1 and G_2 must be "1". The bottom panel of fig. 1.13 shows another logical gate, called OR. This gate has an output of "1" if either G_1 OR G_2 are "1". The last gate we need in order to make an adder is the so-called *exclusive OR* or XOR.

It takes a bit of patience, but it can be verified that the combination of logical gates in fig. 1.15 performs the addition of two binary numbers including the carry-in and carry-out.

How does a computer multiply two numbers? A multiplication between two numbers x and y means that we add the number x, for example, to itself y times. Equivalently, you can think of adding the number y to itself x times and the result is the same. For example, $3 \times 4 = 12$ can be calculated by doing $(3 + 3)$ 4 times, i.e. $3 + 3 + 3 + 3$ or $(4 + 4)$ 3 times, i.e. $4 + 4 + 4$. Both ways we get the correct answer, 12.

A more interesting question is "how does a computer divide two numbers?" While multiplication can be broken down to a series of additions, a division can be broken down into a series of subtractions. Suppose we want to calculate the ratio $\frac{13}{2}$. The calculation proceeds as follows: you subtract the denominator from the numerator and check if what you obtain is bigger than the denominator. If the answer is "yes", then you add 1 to a counter (set to zero initially). If the answer is no, then the result of the division is the value of the counter and the rest is the number obtained before the last subtraction. In

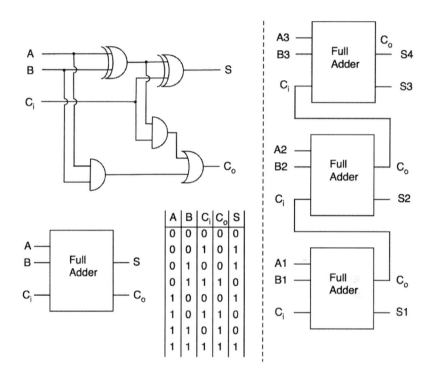

FIGURE 1.15 Left: Schematics of a 1-bit full adder, i.e. the addition takes into account the carry input C_i and produces the output S with the carry output C_o. The table describing the operations and a symbol for the full adder is also shown. Right: 3 adder blocks are connected to produce a 3-bit adder. This circuit adds A1A2A3 to B1B2B3 to produce the output S1S2S3S4, which includes the carryover.

the case of $\frac{13}{2}$, we set the counter $c = 0$ and subtract 2 from 13. The result is 11 and is bigger than 2, so add 1 to the counter. Now subtract 2 from 11 with the result $= 9$ still bigger than 2. So add 1 to the counter. You discover that you reach the number 1, less than two, after 6 subtractions. So the result is $\frac{13}{2} = 6$ with rest 1.

Now that we know how a computer or a pocket calculator can perform the 4 operations, we can now understand how the inverse sine function can be calculated. This is achieved by identifying a so-called *series expansion* , i.e. a summation of usually infinite terms that better and better approximate the value of our function, at a specific point, the more terms we add.

1.3.1 Functions

We have seen that in order to progress with Eratosthenes' calculation we had to define the function sine. But, what is a function? We can think of a function as a "recipe" to calculate a number, given another number. For example, the recipe "calculate the square" is a function that, given a number in input, spits out its square. Feed 2 and spits out 4 or feed 8 and spits out 64, and so on. In a Cartesian plane x, y, we can represent such a function with a plot (see fig. 1.16). We can build such a plot because we can build couples of numbers (x, y) where $y = x^2$.

Usually a function is indicated with the notation $y = f(x)$, which means that the value y is a function (that's why the letter f) of the value x. x is indicated as the *independent variable*, meaning that x is the input value and we can give any number we wish. y is the result of the calculation and is therefore called the *dependent variable* because it is normally uniquely identified by the "recipe". In the case of fig. 1.16, the recipe is "input x, square it and assign the output to y", then build the plot by putting a dot at each (x, y) position in the Cartesian coordinate plane.

1.3.2 Infinity in Maths

Zeno (see fig. 1.17) was a remarkable Greek philosopher living around 450 BC. His teacher, Parmenides, was another Greek philosopher with an interesting view of reality, i.e. reality is just one timeless thing where changes are impossible. As a consequence, in the Universe, nothing moves and the motion we perceive is an illusion. We can assume that people found Parmenides' theories about motion a bit difficult to accept in view of the fact that motion is everywhere as anybody can easily see. Zeno wanted to give substance to Parmenides' claim by proposing four famous paradoxes showing that motion is an illusion or, at least, has some logical problem. Zeno's first paradox (see fig. 1.18) states that you cannot walk from point A to point B because before reaching point B, you need to cross half-distance $\overline{AB}/2$, then half of the remaining distance $\overline{AB}/4$, then half of the remaining distance $\overline{AB}/8$ and so on for an infinite number of steps, smaller and smaller, but infinite in number.

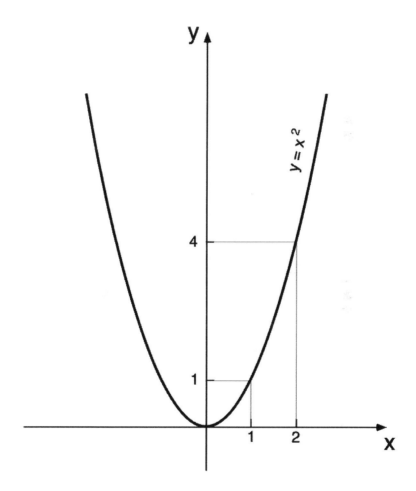

FIGURE 1.16 Plot of the function $y = x^2$.

FIGURE 1.17 Zeno of Elea (495-430 BC) was a Greek philosopher known mostly for his subtle paradoxes.

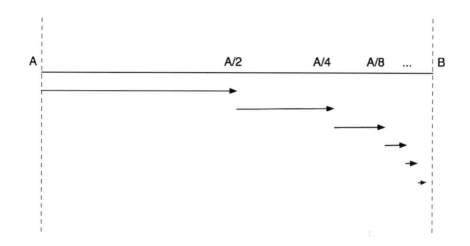

FIGURE 1.18 Zeno's first paradox. Starting from point A point B cannot be reached because before reaching B, you need to cross half-distance, then half of the remaining distance, and so on. No matter how many steps you take you never reach point B (or after an *infinite* number of steps, not acceptable at Zeno's time).

Zeno then argues that, no matter how many steps you take, you will always be between A and B and incapable of reaching B.

You can imagine that such argument is very powerful and for many centuries people could not refute it easily.

With a modern view, from Zeno's paradox we learn something extremely important: *it is possible to divide a segment into an infinite number of smaller and smaller segments. Or, alternatively, it is possible to sum an infinite number of numbers and end up with a finite number.* In the case of Zeno's paradox, we just showed that:

$$\overline{AB} = \frac{\overline{AB}}{2} + \frac{\overline{AB}}{4} + \frac{\overline{AB}}{8} + \frac{\overline{AB}}{16} + ... \tag{1.11}$$

$$\overline{AB} = \overline{AB}(\frac{1}{2} + \frac{1}{4} + \frac{1}{8} + \frac{1}{16} + ...) \tag{1.12}$$

$$1 = \frac{1}{2} + \frac{1}{4} + \frac{1}{8} + \frac{1}{16} + ... \tag{1.13}$$

The number 1 can be obtained by adding an infinite number of smaller and smaller fractions. There is a better way to write the last line of eq. 1.13 by using the symbol $\sum_{n=1}^{\infty}$ to indicate a summation of terms in which the exponent n runs from 1 to ∞.

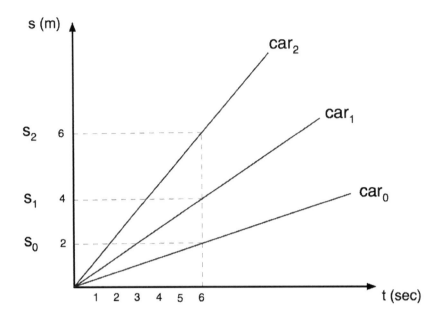

FIGURE 1.19 Three cars running at different constant speed.

$$\sum_{n=1}^{\infty} \frac{1}{2^n} = \frac{1}{2} + \frac{1}{4} + \frac{1}{8} + \frac{1}{16} + \ldots = 1 \qquad (1.14)$$

It is important to notice that the series 1.14 belongs to the special class of series that are said to be *convergent*, i.e. the summation converges to a finite number. Many series do not converge and we refer the reader to specialized books to study the convergence criteria of infinite series.

We are now equipped to show how to calculate velocities and areas and show that these two operations are related.

1.3.3 Derivatives, Integrals and the Fundamental Theorem of Calculus

The study of physics concerns, very often, the study of how systems evolve with time. For example, the orbital motion of a planet is well understood if we can predict with some accuracy its position at a certain given time. It is very important, therefore, to study how objects move in space and time and what maths is appropriate to study time evolution. Let's study the motion of cars cruising at constant speed. We can draw a Cartesian coordinate system where the horizontal axis is time and the vertical axis is space. Suppose that three cars are traveling on a straight line, each with a different constant speed.

This means that at a certain time 6, for example, car_0 will be in the position s_0, car_1 in the position s_1, and car_2 in position s_2. How does a plot of these positions look on a graph? In fig. 1.19 we notice that first, each car traces a straight line in the graph; second, all the lines intersect at the point (0,0) and third, the faster the car the more inclined the straight line is. For example, car_2 is faster than car_1 because in the same time interval of 6 seconds, car_2 has run a distance of 6 meters compared to the distance of 4 meters run by car_1. It is easy now to see that car_1 is faster than car_0. A car that does not move will have a straight line coincident with the t-axis: as the time passes, the car is at 0 distance. A car travelling at infinite speed will have a straight line coincident with the s-axis, i.e. in zero time is at infinity!

There is clearly a relationship between how inclined the lines are and how fast a car is going. It is a simple fact that the speed of a car is measured in km/hour or miles/hour. The way to measure the speed is to calculate what distance has been run by a car in a given time interval. If we use meters and seconds for, respectively, the measurements of distances and time, we measure speed in meters/second. With reference to fig. 1.20, we can see that our car is moving from O to D in the time taken to go from O to A. So the segment $\overline{OD}$ measures the space travelled by the car in meters while the segment $\overline{OA}$ measures the time needed to go from O to D. The speed of the car is simply the ratio of the two segments $\frac{\overline{OD}}{\overline{OA}}$. We also notice that the segment $\overline{OD}$ is proportional to the sine of the angle α, while the segment $\overline{OA}$ is proportional to the cosine of the angle α. We therefore have that the speed of the car can be expressed geometrically as the ratio:

$$speed = \frac{\sin \alpha}{\cos \alpha} = \tan \alpha \qquad (1.15)$$

where a new trigonometric function, *tangent* of α ($\tan \alpha$), has been introduced[3]. In the motion of the car, we can also say that the space is a function of time, i.e. given a time t we can calculate the space t. In order to do so, we need to know the speed of the car which is:

$$speed = \frac{space}{time} = \tan \alpha \qquad (1.16)$$

What is the function that describes the position of the car with respect to time? The function is:

$$s = k \cdot t \qquad (1.17)$$

where k is the speed in m/sec. Putting together equations 1.15, 1.16 and 1.17, we have that the speed is:

$$s = k \cdot t = \frac{\sin \alpha}{\cos \alpha} \cdot t = \tan \alpha \cdot t \qquad (1.18)$$

[3]The speed contains the ratio of $\sin \alpha$ and $\cos \alpha$ and the proportionality factors cancel out.

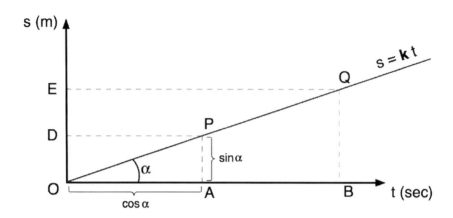

FIGURE 1.20 Geometrical interpretation of velocity.

We have the important result that the speed of the car is equal to the tangent of the angle made by the line representing the motion of the car in a time/space diagram. Another important concept concerns the fact that the speed represents the **variation** of space with respect to **time**, i.e. how space is covered with passing time. If you cover more space within the same interval of time, then you are going faster.

The case of a car cruising at constant speed was easily treated. A more difficult problem would be to calculate the speed of a car that *does not* cruise at constant speed. In this case the plot of the motion of the car is more complex. Let's consider, for simplicity, the case in which the function describing the motion is quadratic with respect to time, i.e. doubling the time makes 4 times the space, etc.:

$$s = t^2 \tag{1.19}$$

It is evident that the car is now moving with a motion such that the speed changes constantly. So now, in addition to having **variations of space with time** we also have **variations of speed with time**.

With the speed changing constantly it is now difficult to calculate the speed of the car. We can calculate the *average* speed by taking two positions in space whose distance is Δs and calculate the time difference Δt. The average speed will be $v_{avg} = \Delta s / \Delta t$.

We can be a bit bold and ask whether it is possible to calculate the speed at any time. In other words, given a time t, I know what space is covered by the car $s = t^2$, but can I calculate the **instantaneous** velocity? Is there an expression, or a function that given how space changes with time, gives

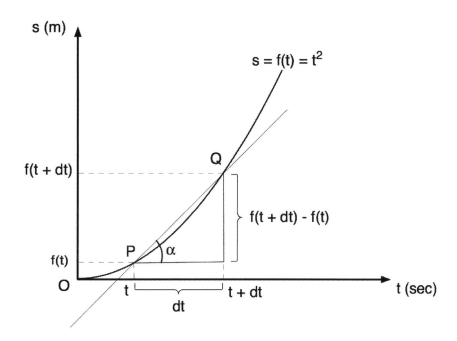

FIGURE 1.21 Calculation of average speed between t and t+dt. If dt becomes smaller and smaller, the average speed tends closer and closer to the *instantaneous* speed at P.

how speed changes with time? The answer is yes and the operation is called a *derivative*.

Let's now look at fig. 1.21. We want to be able to calculate the speed of the car at exactly a point P. We can approximate the speed by calculating the average speed between point P and a point Q very very close to P. If we trace a straight line between P and Q then we know how to calculate the speed. It is the *tan* α, i.e. the ratio of the segment f(t+dt) - f(t), i.e. the sin α, with the length of the segment (t+dt) - dt = dt, i.e. the cos α. The closer to P we choose the point Q, the better the estimation of the instantaneous speed at P will be. If we make **infinitesimally** small the time interval dt, we are **infinitesimally** close to the instantaneous speed. We define the derivative of the function f(t):

$$\frac{df}{dt} = \lim_{dt \to 0} \frac{f(t+dt) - f(t)}{dt} \qquad (1.20)$$

The symbol limit $\lim_{dt \to 0}$ means that we are allowing the value of dt to get closer and closer to zero. We immediately recognize the tangent of α after the symbol *lim*. We now have a recipe to find the derivative of a function, in our case the instantaneous velocity of the car. If we plug into eq. 1.20 the function describing the non-uniform motion of the car $f(t) = t^2$, we have:

$$
\begin{aligned}
\frac{df}{dt} = \lim_{dt \to 0} \frac{f(t + dt) - f(t)}{dt} &= \lim_{dt \to 0} \frac{(t + dt)^2 - (t^2)}{dt} \\
&= \lim_{dt \to 0} \frac{t^2 + 2tdt + dt^2 - t^2}{dt} \\
&= \lim_{dt \to 0} \frac{dt(2t + dt)}{dt} \\
&= \lim_{dt \to 0} (2t + dt)
\end{aligned}
\tag{1.21}
$$

In the last equation we can safely allow dt to be exactly zero. It follows that:

$$
\begin{aligned}
f(t) &= t^2 \\
v = \frac{df}{dt} &= 2t
\end{aligned}
\tag{1.22}
$$

There is a straightforward visualization of the derivative of the function $f(t) = t^2$. In fig. 1.22 the function $f(t) = t^2$ is represented as a square of side t. If each side is increased by dt, the total area increase will be $tdt + tdt + dt^2 \approx 2tdt$. We neglect the term dt^2 because it is the product of two infinitesimals that can be ignored to the first order. The area change per unit dt is therefore $2tdt$, which is the derivative of the function t^2.

We have above a very powerful set of equations. If we want to calculate where the car is at time $= 2$ seconds, we plug the number 2 into the first equation in 1.22. But now if we want to know the instantaneous speed at time $t = 2$ seconds, we plug the number 2 into the second of the equations 1.22. If we do this, we obtain that the car has run for s $= 4$ meters and it has a speed $v = 4$ meters/second. We see that the speed of the car depends on the time t according to the function $f = 2t$.

In general, it can be shown that the derivative of any power of t can be expressed as:

$$
\begin{aligned}
f(t) &= kt^n \\
\frac{df}{dt} &= nkt^{n-1}
\end{aligned}
\tag{1.23}
$$

where k is a constant and n is the power exponent. It can be checked easily that applying eq. 1.23 to the function $f = t^2$ gives the calculated answer. Equation 1.23 allows us to calculate the derivative of polynomials[4] of any

---

[4]A polynomial is a mathematical expression consisting of one or more variables and numerical coefficients. The various terms in a polynomial are only added, subtracted and multiplied while the variables appear only with integer exponents. A generic polynomial in one variable x is $P(x) = a + bx + cx^2 + dx^3 + ...$ where a, b, c and d are constants.

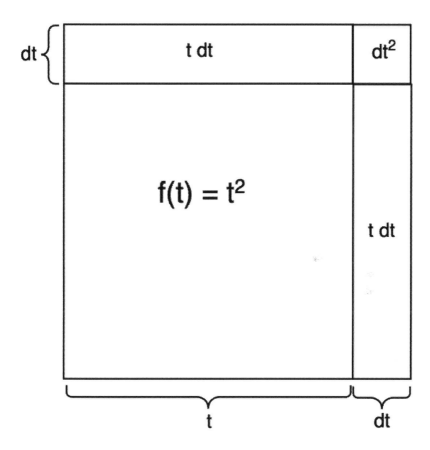

FIGURE 1.22 Graphical representation of the derivative of the function $f(x) = t^2$.

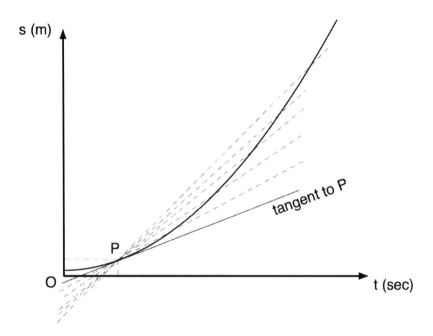

FIGURE 1.23 How to build the tangent line to point P.

Function f(x)	Derivative
x^n	nx^{n-1}
$\sin x$	$\cos x$
$\cos x$	$-\sin x$
e^x	e^x
b^x	$b^x \ln(b)$
$\ln(x)$	$\frac{1}{x}$
$\sin^{-1} x$	$\frac{1}{\sqrt{(1-x^2)}}$
$\cos^{-1} x$	$\frac{-1}{\sqrt{(1-x^2)}}$

TABLE 1.1   Table of derivatives.

degree. It is important to notice that the derivative of any constant is equal to 0 since there is no change in value of the constant with time. Table 1.1 shows a list of calculated derivatives of some common functions.

We have a bonus concept out of our definition of derivative. What happens to the straight line we used to define the average speed over smaller and smaller time intervals? In fig. 1.23 we see how the straight lines approach, closer and closer, the line tangent at the point P. Can we calculate the function defining the tangent line to a point P over a curve?

Yes. First, the tangent is a straight line and the general equation for a straight line is:

$$s = mt + q \tag{1.24}$$

in eq. 1.24, s is the space, t is the time. In order to uniquely define the tangent to a curve, we need the equation of the curve and the coordinates of where to calculate the tangent. Let's call $s = s(t)$ the function and let the point P have coordinates (t_0, s_0). The equation of the tangent is:

$$s - s_0 = \left(\frac{ds}{dt}\right)_{t=t_0} \cdot (t - t_0) \tag{1.25}$$

The symbol $\left(\frac{ds}{dt}\right)_{t=t_0}$ represents the value of the derivative calculated at the point $t = t_0$ and is therefore a number. Notice that equation 1.25 can be written as eq. 1.24 if:

$$m = v = \left(\frac{ds}{dt}\right)_{t=t_0}$$

$$q = s_0 + t_0 \left(\frac{ds}{dt}\right)_{t=t_0} \tag{1.26}$$

where v is the speed (or the modulus of the velocity: we'll see the meaning and difference between *speed* and *velocity* later in the book).

Very often we will encounter functions that are the product of two functions, usually depending on the time t variable. We need a recipe to calculate the derivative of a function that is the product of functions. Given a function $w(t) = u(t) \cdot v(t)$, its derivative is given by:

$$\frac{d}{dt}[w(t)] = \frac{d}{dt}[u(t) \cdot v(t)] = v \cdot \frac{du}{dt} + u \cdot \frac{dv}{dt} \qquad (1.27)$$

We now show how to prove eq. 1.27. We start with the definition of derivative in eq. 1.20:

$$\begin{aligned}\frac{d}{dt}[w(t)] &= \lim_{\Delta t \to 0} \frac{w(t + \Delta t) - w(t)}{\Delta t} \\ &= \lim_{\Delta t \to 0} \frac{u(t + \Delta t)v(t + \Delta t) - u(t)v(t)}{\Delta t}\end{aligned} \qquad (1.28)$$

We now add and subtract the same quantity, $u(t)v(t + \Delta t)$, to the numerator:

$$\begin{aligned}\frac{dw}{dt} &= \lim_{\Delta t \to 0} \frac{u(t + \Delta t)v(t + \Delta t) - u(t)v(t + \Delta t) + u(t)v(t + \Delta t) - u(t)v(t)}{\Delta t} \\ &= \lim_{\Delta t \to 0} \frac{(u(t + \Delta t) - u(t)) \cdot v(t + \Delta t) + u(t) \cdot (v(t + \Delta t) - v(t))}{\Delta t} \\ &= \lim_{\Delta t \to 0} \frac{u(t + \Delta t) - u(t)}{\Delta t} \cdot \lim_{\Delta t \to 0} v(t + \Delta t) \\ &\qquad + \lim_{\Delta t \to 0} v(t) \cdot \lim_{\Delta t \to 0} \frac{v(t + \Delta t) - v(t)}{\Delta t} \\ &= \frac{du}{dt} \cdot v(t) + u(t)\frac{dv}{dt}\end{aligned}$$

$$(1.29)$$

In analogy to the case of the visual representation of the derivative of t^2, we can show visually the origin of the chain rule (see fig. 1.24). The increase in the function is approximately equal to the sum of the area of the two rectangles $udv + vdu$. The term $dudv$ is higher order and can be neglected.

Another useful rule is the derivative of a function of a function, i.e. where we have a function $f = g[h(t)]$. In this case, we have:

$$\frac{d}{dt}g(h(t)) = \frac{dg}{dh}\frac{dh}{dt} \qquad (1.30)$$

This is usually called the *chain rule*[5].

---

[5]The proof of the chain rule is a bit tricky and we refer the reader to any good book on calculus.

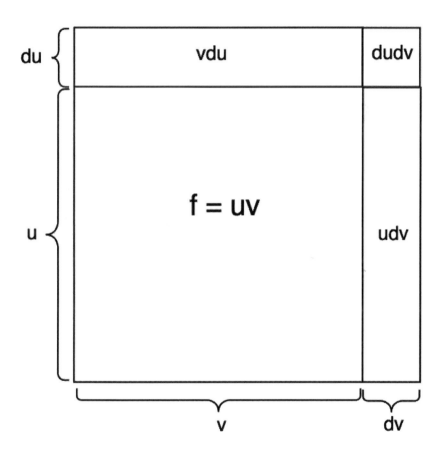

FIGURE 1.24 Visual representation of the derivative of $f = u \cdot v$.

We have seen above that the concept of a derivative of a function is closely connected with the study of how a function changes with respect to time. In the case of the motion of a car, we have studied how to define the instantaneous speed given the law of motion with respect to the variable time t. We would like to ask now the opposite question, i.e. given the knowledge of the speed and how it changes with time, how can we calculate the distance the car has run since some initial time $t = t_0$? The answer will bring us to integral calculus. Notice that this question is the exact opposite of the previous question hinting that integral calculus is somehow the reverse of differential calculus.

We have seen in eq. 1.22 that the speed is the derivative of the position function $s = s(t)$. If we now know instead the function $v = v(t)$, can we obtain the position $s = s(t)$? Yes, and the operation is called *integration*, which is effectively the anti-derivative. Schematically we can write that:

$$f(t) \leftrightarrows \int \frac{df(t)}{dt} dt \tag{1.31}$$

where the symbol $\int$ is the integral operation. Eq. 1.31 represents the *Fundamental Theorem of Calculus* and it states that derivative and integral are opposite operations or, differentiating and then integrating a function $f(t)$ leaves the function unaltered (almost, see later).

It is straightforward to see the geometrical interpretation of the integral. In fig. 1.25, upper right, the diagram of a car running at constant speed is shown where the vertical axis is the speed and horizontal axis is the time. We know that the law of motion tells us the space travelled according to the equation $s = v(t)t$, in general. In the particular case of a constant speed the equation $s = vt$ is telling us that the space travelled is equal to the area under the curve $A(t) = vt$ because the area of the rectangle is exactly $A(t)$.

Let us now study the variation of the function $A = A(t)$ in the small interval between t and $t + \delta$. The area under the curve between t and $t + \delta$ is equal to the space travelled by the car in the small time interval δ. This space can be written as:

$$|A(t + \delta) - A(t)| = v(t) \cdot \delta - \epsilon \tag{1.32}$$

where ϵ is the error in the area estimation due to the fact that the speed is changing in the little interval δ. If we want this error to be smaller and smaller, we just need to have δ smaller and smaller. We see that:

$$v(t) = \frac{A(t + \delta) - A(t)}{\delta} - \frac{\epsilon}{\delta} \tag{1.33}$$

If we now assume that the ratio $\frac{\epsilon}{\delta} \to 0$ as $\delta \to 0$ [6] we have:

---

[6] It can be shown that the excess error ϵ is less or at least equal to the rectangle $(v(t + \delta) - v(t)) \cdot \delta$. Therefore we can write that $\frac{\epsilon}{\delta} \leq \frac{(v(t+\delta) - v(t)) \cdot \delta}{\delta} = v(t + \delta) - v(t)$ which goes to 0 when δ goes to zero (for a continuous function).

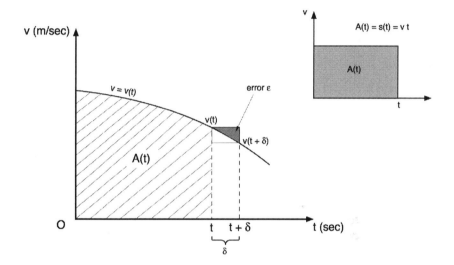

FIGURE 1.25 Geometric interpretation of the integration operation. The upper right diagram shows that the displacement of a car running at constant speed is represented by the area under the line $v =$ constant. The area $A(t)$ is the product of the constant speed with the time t. The lower left diagram shows the general case where the speed is a function of time $v = v(t)$. Also in this case the area $A(t)$ represents the instantaneous distance run by the car at the time t.

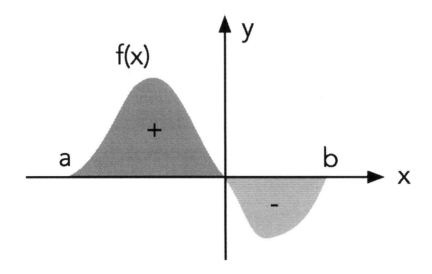

FIGURE 1.26 Interpretation of the definite integration operation as area subtended under the function $f(x)$. Note that the area is defined positive when contained in the $y = f(x) > 0$ semi-plane and negative when contained in the $y = f(x) < 0$ semi-plane.

$$v(t) = \lim_{\delta \to 0} \frac{A(t+\delta) - A(t)}{\delta} = \frac{dA}{dt} \qquad (1.34)$$

From eq. 1.20 we see that eq. 1.34 is telling us that the derivative of the function $A(t)$ is the function $v(t)$ or, in other words, the function $A(t)$ is the *anti-derivative* of the function $v(t)$. We can therefore write that:

$$A(t) = \int v(t')dt' \qquad (1.35)$$

So, the operation of integration of a function $f(x)$ of a variable x, is associated with the determination of the area under the curve represented by the function $f(x)$ in Cartesian coordinate (see fig. 1.26). More in general, the integral operation associates numbers to the summation of infinitesimal quantities and therefore can be associated with physical quantities like, for example, displacement, as well as geometrical quantities like areas, volumes, etc. When we want to integrate a function $f(x)$ over a specific interval $a < x < b$, we write:

$$F = \int_a^b f(x)dx \qquad (1.36)$$

where F is a number. Notice that with our definition the calculated area F

can be a positive as well as negative number. For example, in fig. 1.26, we see that when $f(x)$ is above the line $y = 0$ the area is ≥ 0 while when $f(x)$ is below the line $y = 0$, the area is ≤ 0. If $F(x)$ is a function such that $F(x) = \frac{df}{dx}$ then we have a recipe to calculate the definite integral:

$$\int_a^b f(x)dx = F(x)|_a^b = F(b) - F(a) \tag{1.37}$$

1.3.4 Inverse Trigonometric Functions

Now that we have the techniques of infinitesimal calculus under our belt we can approach the problem of calculating inverse trigonometric functions, i.e. how to calculate the value of the angle whose sine (or cosine or tangent, etc.) is a given number. In the case of Eratosthenes , we have that the sine of the angle we need is the ratio of the lengths of two segments. There is a simple device that does that for you and it is called a **protractor**. We can imagine that Eratosthenes used such a device. Such a tool is good when the accuracy requested is not particularly important. If we want to achieve higher and higher accuracy, we need to find an approximate series that gives us the inverse sine function, i.e. gives us the angle when we know its sine.

We have seen that a computer can easily calculate polynomials. What if we can *approximate* functions with polynomials? Perhaps we can have polynomials with infinite terms and we just need to add enough terms to reach the precision that we need.

It turns out that we can approximate functions with polynomials. A generic polynomial can be written as:

$$P(x) = a_0 + a_1 x + a_2 x^2 + a_3 x^3 + ... = \sum_{n=0}^{\infty} a_n x^n \tag{1.38}$$

The idea is to express *any* function as a series like eq. 1.38. There is a beautiful relationship between the coefficients a_n in eq. 1.38 and the derivatives of the function $P(x)$. For polynomials, it is easy to verify that:

$$a_0 = P(0)$$
$$a_1 = \left(\frac{dP}{dx}\right)_{x=0}$$
$$a_2 = \frac{1}{2}\left(\frac{d^2 P}{dx^2}\right)_{x=0} \tag{1.39}$$
$$a_3 = \frac{1}{2\cdot 3}\left(\frac{d^3 P}{dx^3}\right)_{x=0}$$
$$...$$

For example, let's check that the polynomial $P(x) = 3 + 5x + 3x^2 - 8x^3$ satisfies eq. 1.39:

$$a_0 = P(0) = 3$$

$$a_1 = \left(\frac{dP}{dx}\right)_{x=0} = 5 + 6x - 24x^2 = 5$$

$$a_2 = \frac{1}{2}\left(\frac{d^2P}{dx^2}\right)_{x=0} = \frac{1}{2}(6 - 48x) = 3 \qquad (1.40)$$

$$a_3 = \frac{1}{2\cdot 3}\left(\frac{d^3P}{dx^3}\right)_{x=0} = \frac{1}{6}(-48) = -8$$

so, polynomials can be written as:

$$P(x) = a_0 + a_1 x + a_2 x^2 + a_3 x^3 + \ldots$$

$$= P(0) + \left(\frac{dP}{dx}\right)_{x=0} + \frac{1}{2}\left(\frac{d^2P}{dx^2}\right)_{x=0} + \frac{1}{2\cdot 3}\left(\frac{d^3P}{dx^3}\right)_{x=0} + \ldots \qquad (1.41)$$

A compact way to express eq. 1.41 is:

$$f(x) = f(0) + \frac{1}{n!}\sum_{n=1}^{\infty}\left(\frac{d^n f}{dx^n}\right)_{x=0} x^n \qquad (1.42)$$

where $n! = 1\cdot 2\cdot 3\cdot 4\cdot \ldots \cdot n$ and is called *factorial* of the number n.

The important result is that, if a function $f(x)$ can be (infinitely) differentiated, then we can apply eq. 1.42 and express it as an infinite series of powers. This will allow any computer to calculate any of these functions at any point where the function is defined. Can we express the sine and cosine as power series? Yes. The expansions are given below:

$$\sin x = \sum_{n=0}^{\infty}(-1)^n\frac{x^{2n+1}}{(2n+1)!} \approx x - \frac{1}{3!}x^3 + \frac{1}{5!}x^5 - \ldots \qquad (1.43)$$

$$\cos x = \sum_{n=0}^{\infty}(-1)^n\frac{x^{2n}}{(2n)!} \approx 1 - \frac{1}{2!}x^2 + \frac{1}{4!}x^4 - \ldots \qquad (1.44)$$

The inverse trigonometric functions, that we need to calculate the Eratosthenes angle, are:

$$\sin^{-1}x = \sum_{n=0}^{\infty}\frac{(2n)!}{4^n(n!)^2}\frac{x^{2n+1}}{(2n+1)} \approx x + \frac{1}{2}\frac{x^3}{3} + \frac{1\cdot 3}{2\cdot 4}\frac{x^5}{5} + \ldots \qquad (1.45)$$

$$\cos^{-1}x = \frac{\pi}{2} - \sin^{-1}x \qquad (1.46)$$

We finally have an expression in terms of power series of the function $sin^{-1}x$. Remember that Eratosthenes obtained that $x = \sin\alpha = 0.1256$. By using the first power series in eq. 1.46 we can write that:

$$
\begin{aligned}
\alpha &\approx \frac{180}{\pi}\left(x + \frac{1}{2}\frac{x^3}{3} + \frac{1\cdot 3}{2\cdot 4}\frac{x^5}{5} + ...\right)\\
&= \frac{180}{\pi}\left(x0.1256 + \frac{1}{2}\frac{(0.1256)^3}{3} + \frac{3}{6}\frac{(0.1256)^5}{5} + ...\right) \quad (1.47)\\
&= \frac{180}{\pi}(0.1256 + 0.000330231 + 0.000003126 + ...)\\
&= 7.2154°
\end{aligned}
$$

where the prefactor $180/\pi$ gives the answer in degrees. A pocket calculator with inverse sine function will evaluate the angle $\alpha = 7.2154°$. The sum of the first three terms in eq. 1.47 will also provide $\alpha = 7.2154°$.

1.4 ERATOSTHENES'S FINAL CALCULATION

Now we have all the ingredients to finally estimate the circumference (and therefore the radius) of the Earth. Eratosthenes correctly measured the angle α (see fig. 1.11). So, if the angle α corresponds to the distance between Alexandria and Syene[7], then when $\alpha = 360°$, the corresponding length is the circumference of the Earth . Or, alternatively we can say that the ratio of the distance between Alexandria and Syene to the angle α is equal to the ratio of the circumference of the Earth to the full angle of $360°$ or 2π radians. It follows that the circumference of the Earth must be equal to $\frac{2\pi}{\alpha}$ times the distance from Alexandria to Syene. Let's call D the distance between Syene and Alexandria and C the circumference of the Earth we are trying to estimate. We have:

$$\frac{C}{D} = \frac{2\pi}{\alpha} \quad (1.48)$$

Eratosthenes's tools most probably did not allow him to have a better precision for α than $0.1°$. Therefore he measured $\alpha = 7.2°$. It is now easy to estimate the circumference of the Earth C by inverting eq. 1.48 to extract C. First notice that the ratio of the full angle $360°$ to the measured angle $\alpha = 7.2°$ is exactly equal to 50. This means that the full circumference of the Earth is 50 times the distance from Alexandria to Syene. Eratosthenes knew the distance in *stadia* between the two cities $D = 5000$ stadia. Unfortunately we do not have a value for such an ancient distance unit. If we assume that a stadium is equal to 184.8 meters, as indicated by most of the experts, we have the amazing result that the circumference of the Earth measured by

[7]We are assuming, as Eratosthenes did, that the arc connecting Alexandria and Syene is a great circle, i.e. the largest circle that can be drawn on a sphere.

Eratosthenes was $C = 46,000$ km, or about 15% bigger than today's accepted value.

1.5 ARISTARCHUS'S CALCULATIONS

We conclude this chapter with another remarkable estimate of astronomical quantities by the Greeks and based on simple observations of the Earth-Moon-Sun system. There are a number of simple facts that we are all familiar with: the Moon goes through phases, i.e. the surface of the Moon visible from Earth is illuminated by the Sun and the shape of the portion illuminated directly changes with time. The phases go from no illumination (New Moon) to fully illuminated (Full Moon). There are also special times in which the Moon completely covers the Sun (total solar eclipse) or it enters the shadow of the Earth (lunar eclipse).

Aristarchus of Samos (see fig. 1.27) was a Greek mathematician and philosopher known mostly for being the first to propose that the Sun is at the center of the solar system well ahead of Copernicus . In his book *On the Sizes and Distances of the Sun and Moon*, he discussed how he obtained the ratio of distances from the Earth to the Moon and from the Sun to the Earth. Aristarchus was aware of the phases of the Moon, and he used them to perform his calculations. There is a special day when the face of the Moon appears exactly half illuminated as seen from the Earth (see fig. 1.28). When this happens, it means that the angle Sun-Moon-Earth is exactly 90°. If now we measure the angle θ Moon-Earth-Sun, we know that this angle is defined as the ratio of the segments D_{EM} (distance Earth–Moon) over D_{ES} (distance Earth–Sun). We have:

$$\cos\theta = \frac{D_{EM}}{D_{ES}} = \frac{1}{390} \tag{1.49}$$

Aristarchus measured the angle $\theta = 87°$ which corresponds to a distance from the Earth to the Sun about 20 times the distance from the Earth to the Moon. The reasoning was correct but his measurements were inaccurate. Today we know that the angle is $\theta = 89.853°$ very close to 90°, thus putting the Sun about 390 times the distance from the Earth to the Moon away.

Aristarchus made another observation: he noticed that a total solar eclipse, i.e. when the disk of the Moon exactly covers the Sun, lasted only a few minutes. This means that the apparent angular diameter of the Sun and the Moon must be very close to each other. But Aristarchus already estimated that the Sun is about 20 times farther away than the Moon and so it must be 20 times *bigger* than the Moon. This conclusion is a consequence of the definition of the tangent of an angle as a ratio of two segments as in fig. 1.29. In this figure, the observer is at the point E, the Moon radius is $R_M = \overline{AM}$ and the Sun radius is $R_S = \overline{CS}$. From the figure it is evident that $\tan\alpha$ is defined equivalently as the ratio of $\frac{AM}{EM}$ or $\frac{CS}{ES}$, where $D_{EM} = \overline{EM}$ and $D_{ES} = \overline{ES}$ are respectively the distance from the Earth to the Moon and from the Earth

FIGURE 1.27 Aristarchus of Samo (310-230 BC) was a Greek philosopher and astronomer. He was the first to propose that the Sun is at the center of the solar system.

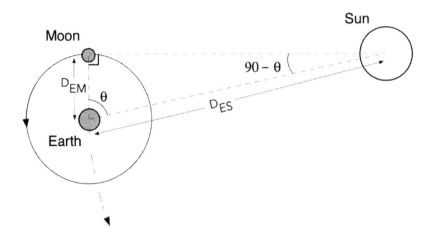

FIGURE 1.28 Geometry used by Aristarchus of Samos to estimate the distance from the Sun to the Earth in units of the distance from the Earth to the Moon.

to the Sun while d_S and d_M are respectively the diameter of the Sun and the Moon.

It follows that:

$$\frac{\overline{AM}}{\overline{EM}} = \frac{\overline{CS}}{\overline{ES}} \tag{1.50}$$

So, the radii of the Sun and the Moon are in the same ratio as their distances from the Earth as Aristarchus pointed out. If we use diameters instead of radii and we use the modern values, we have that the ratio of the diameter of the Sun (d_S) to the diameter of the Moon (d_M) is:

$$\frac{d_S}{d_M} = 390 \tag{1.51}$$

Or equivalently:

$$\frac{d_S}{D_{ES}} = \frac{d_M}{D_{EM}} \tag{1.52}$$

Aristarchus then turned his attention to the observation of the lunar eclipse, i.e. when the Moon enters the shadow of the Earth.

In fig. 1.30 the eclipse geometry used by Aristarchus is shown. In the upper panel we see how, during a lunar eclipse, the Earth projects its shadow in space. When the conditions are right, the orbital plane of the Moon is such that it enters and exits the Earth's shadow. Let's study the medium and bottom panel of fig. 1.30. The two right triangles 1 and 2 are similar. This

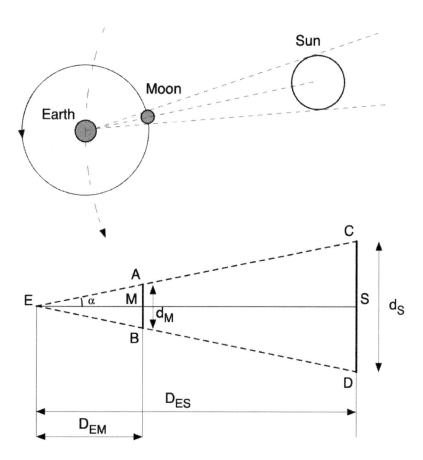

FIGURE 1.29 Geometry of a total solar eclipse used by Aristarchus of Samos to estimate the diameter of the Sun. The bottom panel shows the perspective from an observer on the Earth at point E when the Moon disk AB completely covers the Sun disk CD.

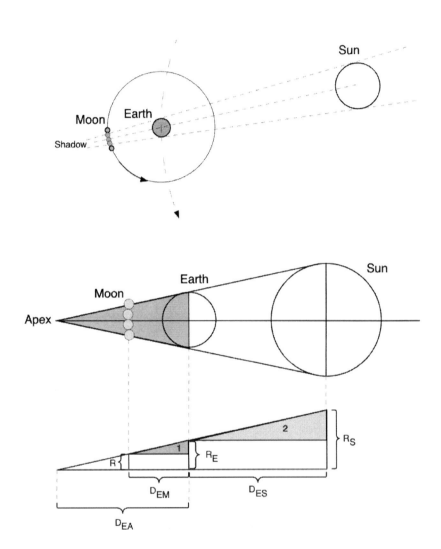

FIGURE 1.30 Geometry used by Aristarchus of Samos to estimate various distances in the system Sun-Earth-Moon.

means that we can identify a number of proportions between pairs of segments of the two triangles. Let's call R the half-length of the Earth's shadow on the Moon trajectory, R_E the radius of the Earth, R_S the radius of the Sun; D_{EM} the distance from the Earth to the Moon, D_{ES} the distance from the Sun to the Earth and D_{EA} the distance between the Moon and the apex of the conical shadow of the Earth in space. The right triangle 1 has, as the short side leg (or *cathetus*), the quantity $(R_E - R)$, and the corresponding short leg of right triangle 2 is equal to $(R_s - R_E)$. The long legs of the two right angles are D_{EM} and D_{ES} for respectively triangle 1 and 2. Due to the similarity, we can write:

$$\frac{D_{EM}}{(R_E - R)} = \frac{D_{ES}}{(R_S - R_E)} \tag{1.53}$$

Now, we know from eq. 1.50 that the distance Earth-Moon and Earth-Sun are in the same ratio as their radii R_M and R_S, where R_M is the radius of the Moon. Plugging eq. 1.50 into eq. 1.53, we have:

$$\frac{D_{ES}}{D_{EM}} = \frac{(R_S - R_E)}{(R_E - R)} = \frac{R_S}{R_M} \tag{1.54}$$

With a little algebra (left as an exercise) we can rewrite eq. 1.54 as:

$$1 + \frac{R}{R_M} = \frac{R_E}{R_M} + \frac{R_E}{R_S} \tag{1.55}$$

Aristarchus observed a lunar eclipse and determined that the radius of the shadow R was equal to twice the radius of the Moon R_M. A modern value is $R = 2.6 \cdot R_M$. With some more algebra we can rewrite eq. 1.54 to express the ratio of the radius (diameter) of the Earth to the radius (diameter) of the Moon:

$$\frac{R_E}{R_M} = \frac{d_E}{d_M} = \frac{1 + \frac{R}{R_M}}{1 + \frac{R_M}{R_S}} \simeq 3.6 \tag{1.56}$$

and we can rewrite eq. 1.54 to express the same ratios for the Sun:

$$\frac{R_S}{R_E} = \frac{d_S}{d_M} = \frac{1 + \frac{R}{R_M}}{1 + \frac{R_S}{R_M}} \simeq 109 \tag{1.57}$$

Aristarchus was able to express all the ratios in terms of the radius of the Earth by looking at the angular diameter of the full Moon.

In fig. 1.31 an observer on the Earth, looking at the full Moon, measures a diameter of the disk equal to 0.519° (modern value). Such a small angle allows us to approximate the diameter d_M with the segment of arc of radius D_{EM}. Under this approximation we can write the proportion:

$$\frac{0.519}{360} = \frac{d_M}{2\pi D_{EM}} \tag{1.58}$$

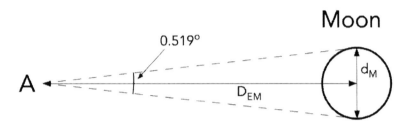

FIGURE 1.31 Geometry to estimate the angular diameter of the Moon.

or

$$\frac{d_M}{D_{EM}} = 9 \cdot 10^{-3} \tag{1.59}$$

giving the ratio between the diameter of the Moon to the distance from the Earth to the Moon. We can also express the distance from the Earth to the Sun in terms of the diameter of the Earth. Let's write the ratio of the distance from the Earth to the Sun D_{ES}, to the diameter of the Earth d_E:

$$\frac{D_{ES}}{d_E} = \frac{d_S}{d_E}\frac{D_{ES}}{d_S} \tag{1.60}$$

Recall from eq. 1.52 that $\frac{d_S}{D_{ES}} = \frac{d_M}{D_{EM}} \rightarrow \frac{D_{ES}}{d_S} = \frac{D_{EM}}{d_M}$. Eq. 1.60 becomes:

$$\frac{D_{ES}}{d_E} = \frac{d_S}{d_E}\frac{D_{EM}}{d_M} \tag{1.61}$$

Using the values calculated in eq. 1.59 and eq. 1.57, we have:

$$\frac{D_{ES}}{d_E} \simeq 1.2 \cdot 10^4 \tag{1.62}$$

Finally, let's calculate the distance from the Earth to the Moon in terms of Earth diameter.

$$\frac{D_{EM}}{d_E} = \frac{1}{\frac{d_E}{D_{EM}}} = \frac{1}{\frac{d_E}{d_M} \cdot \frac{D_M}{D_{EM}}} \simeq 31 \tag{1.63}$$

In summary, Aristarchus was able to calculate the distance to the Sun and the Moon in terms of the Earth's radius or diameter. But we have seen that Eratosthenes was able to calculate the circumference and thus the radius of the Earth. So the Greeks were able to estimate distances in the solar system without any special instruments but just using a lot of geometrical ingenuity.

Think About It...

Euclidean geometry is one example of synthetic geometries. Synthetic geometry starts from statements given without demonstrations (axioms) defining basic entities like points, lines, etc., and proceeds through propositions about these objects without specifying where the objects are located in space. Analytic geometry uses coordinates to transform the propositions into algebraic formulas.

FURTHER READING

Courant, R., and John, F. (1989). *Introduction to Calculus and Analysis. II.* Springer.

Crowe, J., and Shapiro, A. (1981). *Introduction to Algebra and Trigonometry.* Academic Press.

Dreyer, J.L.E. (1953). *A History of Astronomy from Thales to Kepler.* Dover Publications.

Greenberg, M.J., (2008). *Euclidean and Non-Euclidean Geometries: Development and History.* W.H. Freeman.

Kolman, B., and Hayes-Gill, B. (1998). *Introduction to Digital Electronics.* Elsevier Science.

Riley, K.F., Hobson, M.P., and Bence, S.J. (2006) third edition. *Mathematical Methods for Physics and Engineering.* Cambridge University Press.

Math and Physics Toolkit

CONTENTS

I N chapter 3 we will study the motion of planets in the solar system. In order to understand the physics, i.e. describe how the planets move with time, we need to understand the mathematics used in such description. We caution immediately that the math in this chapter and the next might be challenging and we expect the reader to go over the material presented with particular care making sure that all the concepts are understood and clear before progressing.

2.1 VECTORS

We have seen in chapter 1 that we defined velocity and acceleration in terms of time variation of, respectively, space and velocity. In other words, the velocity is the variation of space with respect to time. The acceleration, in a similar way, can be defined as the variation of velocity with respect to time. Both velocity and acceleration are simple numbers. What happens if the object is not traveling on a straight line? In this case velocity and accelerations cannot be described anymore with simple numbers. There is additional information that we have to give: direction with its associated orientation. We can walk at constant speed on a street whose direction is, for example, North-South but we need to say what orientation, i.e. if we are going from N to S or from S to N. It is customary to say "direction" for a vector implying that also the orientation is specified.

A quantity defined by a magnitude and a direction is called a *vector* and is indicated with a letter with a pointed arrow above it[1]. For example, the force vector is indicated with the symbol $\vec{F}$, telling us that the force is specified by a magnitude and an orientation along a direction.

The length from the base to the tip is the *magnitude* while the *direction* is identified by the orientation of the tip. In the case of velocity , we call speed the magnitude of the velocity vector, i.e. how fast the object is moving. So when a physicist asks about speed, he/she is expecting just a single number. If instead the physicist asks about velocity, then he/she wants to know the magnitude and the direction.

When we consider quantities that are completely defined by a number, as is the case of an amount of space, magnitude of velocity, and acceleration **along a straight line**, then we say that the quantity is a scalar quantity[2]. But the concepts of velocity and acceleration become more complex if the object is not constrained to move along a straight line but is free to move in the 3-dimensional space like, for example, a planet orbiting the Sun. In this case velocity and acceleration must be defined not only by giving the magnitude but also the direction with associated orientation. In other words, magnitude and direction can both change with time. So if we want to calculate the derivative with respect to time, we have to consider that both magnitude and direction can change.

For simplicity, we restrict ourselves to a 2-dimensional plane. In fig. 2.1 a simple example of the displacement vector $\overrightarrow{AB}$ on the plane xy is shown. An object travels along the *direction* indicated by the dotted line for an amount of space equal to the length of the segment $\overline{AB}$ (*magnitude*) and *orientation* given by the arrow meaning that the object started in A and ended in B. This displacement vector is represented in a particular Cartesian coordinate system xy. In this particular coordinate system the vector is completely identified by giving its coordinates, i.e. orthogonal projections[3] of the points A and B to the x and y axes. Let's call $\vec{u}$ the vector $\overrightarrow{AB}$ in fig. 2.1. This vector is completely defined by giving its coordinates $\vec{u} = (u_x, u_y)$ where $\vec{e_x}$ and $\vec{e_y}$ are unit vectors parallel, respectively, to the x and y axes[4]. It is easy to see that:

---

[1]This statement is not exactly true. A finite rotation around an axis can be described by giving a direction (the rotation axis) and a magnitude (the amount of rotation) but it is not a vector. Later in this chapter we will be more accurate.

[2]We need to specify that the straight line has an orientation, so positive velocities are those with the same orientation as the straight line while negative velocities are in the opposite orientation.

[3]By orthogonal projection we mean tracing a perpendicular line.

[4]The unit vectors are also called *base vectors*.

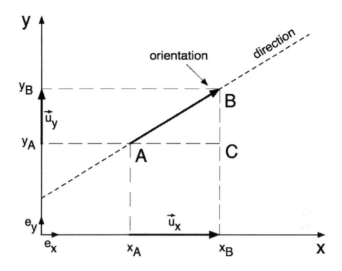

FIGURE 2.1 Displacement vector indicating an object moving from point A to point B (orientation) with a magnitude equal to the length of the segment AB.

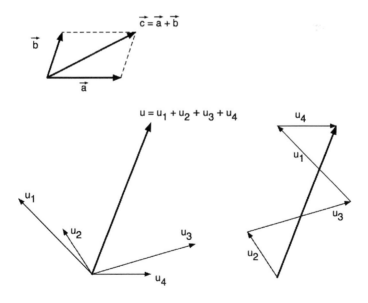

FIGURE 2.2 Geometric addition of two vectors (top) or more vectors (bottom).

$$u_x = (x_B - x_A)$$
$$u_y = (y_B - y_A) \tag{2.1}$$

Now, with reference to fig. 2.1, $x_B - x_A$ is the magnitude of the vector $\vec{u}_x$ while $y_B - y_A$ is the magnitude of the vector $\vec{u}_y$. Notice that by construction the base vectors e_x and e_y belong to the same directions as, respectively, the x and y axes of the coordinate system. This means that we can write:

$$\vec{u} = \vec{e}_x \cdot u_x + \vec{e}_y \cdot u_y \tag{2.2}$$

where we used the geometric rule to add vectors shown in fig. 2.2.

Looking at the triangle $\triangle ACB$, it is easy to verify that the magnitude of the vector $\vec{u}$ can be written as:

$$\| \vec{u} \| = \sqrt{\overline{AC}^2 + \overline{CB}^2} = \sqrt{u_x^2 + u_y^2} \tag{2.3}$$

where the symbol $\|u\|$ indicates the magnitude of the vector $\vec{u}$. The direction is given by:

$$\theta = \arctan\left(\frac{\overline{BC}}{\overline{AC}}\right) = \arctan\left(\frac{u_y}{u_x}\right) \tag{2.4}$$

2.1.1 Change of Coordinate Systems

A vector in a specific coordinate system x, y, z is represented by 3 numbers corresponding to the 3 projections of the vector's tip along the axis. We assume that the base of the vector coincides with the origin of the coordinate system. We can obviously represent the same vector in some other coordinate system in which the vector will be represented by 3 other numbers. This means that the same vector can have different components, i.e. 3 different coordinates, depending on the choice of coordinate system. Therefore we need to be sure that if we change coordinate system, the vector magnitude and direction does not change. This requirement is dictated by requiring that the vector represents a physical quantity and we impose that the physical quantity stays the same no matter what coordinate system we use.

In fig. 2.3, we have two different coordinate systems: one is labeled x, y and the other is labeled x', y'. The primed coordinate system is obtained by shifting it by a quantity a in the x coordinate and a quantity b in the y coordinate. The vector $\vec{u}$ can therefore be projected onto the two different coordinate systems generating two independent sets of coordinates. To simplify the picture, we have indicated the intercepts of $\vec{u}$ to the x' and y' axes by filled dots while the intercepts to the x and y axes are indicated with open dots. The magnitude of the vector $\vec{u}$ can be expressed in the x, y coordinate system by:

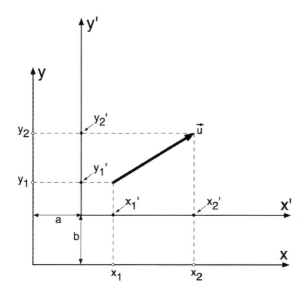

FIGURE 2.3 A vector expressed in two shifted coordinate systems x, y and x', y' maintains the same magnitude and direction.

$$\|\vec{u}\| = \sqrt{(x_2 - x_1)^2 + (y_2 - y_1)^2} \tag{2.5}$$

We know that the primed coordinate x', y' can be expressed as a function of the old coordinate x, y:

$$\begin{aligned} x' &= x - a \\ y' &= y - b \end{aligned} \tag{2.6}$$

The same vector $\vec{u}$ expressed in the primed coordinate system must have the same magnitude. Therefore:

$$\begin{aligned} \|\vec{u}\| &= \sqrt{(x'_2 - x'_1)^2 + (y'_2 - y'_1)^2} \\ &= \sqrt{[(x_2 - a) - (x_1 - a)]^2 + [(y_2 - a) - (y_1 - a)]^2} \\ &= \sqrt{[x_2 - a - x_1 + a]^2 + [y_2 - a - y_1 + a]^2} \\ &= \sqrt{(x_2 - x_1)^2 + (y_2 - y_1)^2} \end{aligned} \tag{2.7}$$

Eq. 2.7 compared with eq. 2.5 shows that the magnitude of the vector is unchanged with a coordinate shift. It is easy to prove that the direction is also unchanged and we leave the proof to the reader.

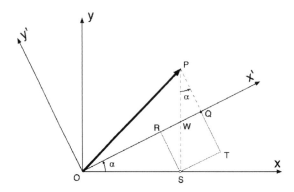

FIGURE 2.4   A vector expressed in two coordinate systems x, y and x', y' rotated by an angle α.

Let's now study the case in which the primed coordinate system is obtained by rotating the original coordinate system by an angle α as shown in fig. 2.4 around a vertical axis (perpendicular to the page) which is also the common origin of the two systems. For simplicity we have assumed that our vector has its base coincident with the origin of the two coordinate systems. The point P is the tip of the vector which has x coordinate equal to the segment $\overline{OS}$ and y coordinate equal to the segment $\overline{PS}$ in the unrotated x, y coordinate system. In the rotated coordinate system x', y', the coordinates of the point P are the segments $\overline{OQ}$ and $\overline{PQ}$ for respectively the x' and y' coordinates. Let's build a rectangle $RSTQ$ where the point R is the projection of the point S onto the x' axis. By construction, the angle $\angle ORS$ is a right angle. The two triangles $\triangle RSW$ and $\triangle PQW$ are similar, i.e. they have all the angles equal[5]. They both have a right angle, and the angle $\angle RWS$ is equal to the angle $\angle PWQ$ because they are opposite. It follows therefore that the angle $\angle RSW$ must be equal to $\angle WPQ$. Now let's turn our attention to the triangles $\triangle OWS$ and $\triangle ORS$. By construction, the angle $\angle ORS$ is a right angle and therefore the angle $\angle RSO$ is equal to $90° - \alpha$, from which we immediately see that the angle $\angle RSW$ must be equal to α. But we have seen that $\angle RSW$ is equal to $\angle WPQ$. It follows that the angle $\angle WPQ = \alpha$. By construction we also see that $\overline{RQ} = \overline{ST}$ and $\overline{RS} = \overline{QT}$.

In the triangle $\triangle ORS$ we see that $\overline{OR} = \overline{OS} \cdot \cos \alpha = x \cdot cos\alpha$. In the triangle $\triangle PST$ we have that $\overline{TS} = \overline{PS} \cdot sin\alpha = y \cdot sin\alpha$. We just found how to obtain the x' rotated coordinate of the point P if we know its x, y coordinates and the angle of rotation α: $x' = x\ cos\alpha + y\ sin\alpha$. In a similar line of reasoning, looking at the triangles $\triangle PST$ and $\triangle ORS$ we see that

---

[5]Two triangles are *similar* if all their angles are equal. In this case all the corresponding sides are in proportion.

$y' = \overline{PQ} = \overline{PT} - \overline{QT} = \overline{PT} - \overline{RS} = y \cdot cos\alpha - x \cdot sin\alpha$. We can collect the two relations we just found in a convenient way:

$$x' = x \ cos\alpha + y \ sin\alpha$$
$$y' = -x \cdot sin\alpha + y \cdot cos\alpha \qquad (2.8)$$

We can look at eq. 2.8 as a recipe to transform the coordinates of a vector, in particular under a rotation of the coordinate system of an angle α. The vector does not change: its representation in different coordinate systems changes according to the recipe given. The vector, somehow, has its own existence independent from the coordinate system that we use to represent it. It is easy to show that the rotation of the coordinate system in eq. 2.8 keeps the magnitude of a vector constant.

The magnitude of a vector written in Cartesian coordinates is given by eq. 2.3. When the base of the vector is coincident with the origin of the coordinate system $(0,0)$, we have:

$$\|u\| = \sqrt{x'^2 + y'^2} = \sqrt{x^2 + y^2} \qquad (2.9)$$

Let us express the primed coordinates using eq. 2.8:

$$x'^2 = x^2 cos^2\alpha + y^2 \sin^2\alpha + 2xy \sin\alpha \cos\alpha$$
$$y'^2 = x^2 sin^2\alpha + y^2 \cos^2\alpha - 2xy \sin\alpha \cos\alpha \qquad (2.10)$$

we now add the two above equations:

$$x'^2 + y'^2 = x^2(\cos^2\alpha + \sin^2\alpha) + y^2(\sin^2\alpha + \cos^2\alpha) = x^2 + y^2 \qquad (2.11)$$

where we used the fact that $(\sin^2\alpha + \cos^2\alpha) = 1$.

We can now formalize a bit more the definition of a vector: a vector is a physical quantity that has a magnitude and a direction *and* its components transform under a rotation according to the recipe given in eq. 2.8. All vectors can be represented by 2 numbers on a 2-dimensional coordinate system. However, not all pairs of numbers represent a vector!

Eq. 2.8 can be written in a different way:

$$\begin{pmatrix} x' \\ y' \end{pmatrix} = \begin{pmatrix} \cos\alpha & \sin\alpha \\ -\sin\alpha & \cos\alpha \end{pmatrix} \begin{pmatrix} x \\ y \end{pmatrix} \qquad (2.12)$$

where the new primed coordinates are indicated as a two-dimensional vector $\begin{pmatrix} x \\ y \end{pmatrix}$, where as usual x, y are the original non-rotated coordinates. The object $A = \begin{pmatrix} a & b \\ c & d \end{pmatrix}$ is called a 2×2 matrix. The rule to multiply a 2×2 matrix with a two-dimensional vector is:

$$A \cdot \vec{u} = \begin{pmatrix} a & b \\ c & d \end{pmatrix} \begin{pmatrix} u_x \\ u_y \end{pmatrix} = \begin{pmatrix} au_x + bu_y \\ cu_x + du_y \end{pmatrix} \tag{2.13}$$

where $\vec{u}$ is a two-component vector $\vec{u} = \begin{pmatrix} u_x \\ u_y \end{pmatrix}$. Eq. 2.13 tells us that the matrix A is an object that acts on a vector to produce another vector. In the specific case of the rotations of eq. 2.12, we can symbolically write:

$$\vec{u'} = A(\alpha)\vec{u} \tag{2.14}$$

where $A(\alpha)$ is given by:

$$A(\alpha) = \begin{pmatrix} \cos(\alpha) & \sin \alpha \\ -\sin \alpha & \cos(\alpha) \end{pmatrix} \tag{2.15}$$

and indicates that the matrix A depends on the rotation angle α. In 3-dimensional space, vectors have 3 components and matrices acting on these vectors are 3×3 tables of numbers.

2.1.2 Operations with Vectors

Having defined vectors we can ask if we can do operations with them. The simplest operation is addition. Can we add two vectors? Yes, and the geometrical recipe is simple and is called the *parallelogram law*. This is a law, meaning that we consider it to be true because it is based on experimental evidence. Various proofs have been given (including Newton) but they are not accepted mathematically. The important fact, which is evidence based, is that forces, among other vectors representing other physical quantities, always obey the parallelogram law. This means that we are justified to use vectors to represent forces, for example, and that experiments have confirmed that where more forces are applied to the same object, they add according to the parallelogram law. That's why it is a **law** and is not proved but *assumed* to be true within the experimental errors.

We have seen that we can add two vectors with the parallelogram rule. We restate the parallelogram rule by saying that the sum of two vectors $\vec{a}$ and $\vec{b}$ is obtained by placing them head to tail and drawing the vector from the free tail to the free head. Alternatively, as shown in fig. 2.2, the sum is obtained by placing the two tails together and drawing two lines (dotted) parallel to each of the vectors. The vector sum $\vec{c}$ is obtained by drawing an arrow with the tail coincident with common tails of the two addends and the head at the intersection of the two dotted lines.

How do we calculate the vector sum of two vectors when they are expressed in a Cartesian coordinate system? It is easy to verify that, given two vectors $\vec{a}$ and $\vec{b}$, of coordinates (a_x, a_y) and (b_x, b_y) in a Cartesian coordinate system x, y, the vector sum $\vec{c} = \vec{a} + \vec{b}$ will be written as:

$$\vec{c} = (a_x + b_x)\vec{e_x} + (a_y + b_y)\vec{e_y} \qquad (2.16)$$

where the recipe to calculate the coordinate of the vector sum is given by equation 2.16. It is easy to see that if we multiply a vector by a number k, the vector does not change direction but it is stretched by an amount equal to k ($\vec{d} = k\vec{c}$). The vector $\vec{d}$ has a magnitude k–times the magnitude of $\vec{c}$ and it will be expressed by:

$$\vec{d} = k \cdot \vec{c} = ka_x\vec{e_x} + ka_y\vec{e_y} \qquad (2.17)$$

and the recipe is simply multiply the components by the stretching constant k. The question is: can we multiply a vector by another vector? We know how to multiply numbers but we really don't know how to multiply things that are not numbers. However, we can come up with a recipe that we call "vector multiplication" that operates on the numerical components of the vector in a certain coordinate system. Vector components are numbers and we know how to multiply them. We actually have two recipes for two different vector multiplications. One recipe takes the components of two vectors and generates a number (or a scalar), as an output. This product is called the *scalar product*[6]. There is another recipe that takes the components and generates a set of components of another vector as an output. This product is called the *vector product*[7] because it generates another vector.

Let us now discuss the origin of the scalar and vector products. The scalar product between two vectors $\vec{a}$ and $\vec{b}$ is indicated by $\vec{a} \cdot \vec{b}$ and produces a number (scalar). We have seen already (eq. 2.3) how to obtain the magnitude of a vector from its coordinates. We have also seen that the magnitude of the vector is independent from the choice of coordinate system. When a quantity does not depend on the choice of the coordinate system we say that the quantity is **invariant** under that specific coordinate transformation. So, the magnitude of a vector is invariant. Finding invariants is a very important and useful thing in physics because if we can express physical laws in terms of vectors, then we are insured that the laws are invariant and we have the freedom to choose whatever coordinate system we want. We will see that certain problems are very difficult to treat in Cartesian coordinates while they become more easily treated in other coordinate systems.

Let's now consider 3 vectors $\vec{a}$, $\vec{b}$ and $\vec{c}$ with coordinates respectively (a_x, a_y), (b_x, b_y) and (c_x, c_y). Taken individually, each vector's magnitude is an invariant:

---

[6]The scalar product is also called the *dot product*. Mathematicians call it the *inner product* or *projection product*.

[7]The vector product is also-called the *cross product*.

$$\|\overrightarrow{a}\|^2 = a_x^2 + a_y^2$$
$$\|\overrightarrow{b}\|^2 = b_x^2 + b_y^2 \qquad (2.18)$$
$$\|\overrightarrow{c}\|^2 = c_x^2 + c_y^2$$

Suppose that the vector $\overrightarrow{c}$ is obtained by summing the two vectors $\overrightarrow{a}$ and $\overrightarrow{b}$, i.e. $\overrightarrow{c} = \overrightarrow{a} + \overrightarrow{b}$. We must have:

$$\|c\|^2 = (a_x + b_x)^2 + (a_y + b_y)^2$$
$$\|c\|^2 = a_x^2 + b_x^2 + 2a_x b_x + a_y^2 + b_y^2 + 2a_y b_y \qquad (2.19)$$
$$\|c\|^2 = \|\overrightarrow{a}\|^2 + \|\overrightarrow{b}\|^2 + 2(a_x b_x + a_y b_y)$$

The last equation in eq. 2.19 tells us that the quantity $(a_x b_x + a_y b_y)$ must be an invariant because all the other terms are invariant. We call this quantity the **scalar product** of $\overrightarrow{a}$ and $\overrightarrow{b}$:

$$\overrightarrow{a} \cdot \overrightarrow{b} = (a_x b_x + a_y b_y) \qquad (2.20)$$

If the vectors live in the 3-dimensional world, then their scalar product is:

$$\overrightarrow{a} \cdot \overrightarrow{b} = (a_x b_x + a_y b_y + a_z b_z) \qquad (2.21)$$

where now we have the extra z axis. From the above definition, it follows that the dot product of a vector with itself is equal to the square of its magnitude:

$$\overrightarrow{a} \cdot \overrightarrow{a} = (a_x^2 + a_y^2) = \|\overrightarrow{a}\|^2 \qquad (2.22)$$

The scalar product always gives a scalar as an output which is invariant under change of coordinate systems[8].

We now show that it is possible to calculate the invariant generated by the dot product *without* knowing the coordinates of the vector but just their magnitudes and directions. Let us consider two vectors $\overrightarrow{a}$ and $\overrightarrow{b}$ as shown in fig. 2.5. We see from the figure that vector $\overrightarrow{a}$ is aligned with the x axis and therefore has coordinates $(a_x, 0)$. The dot product, in this particular Cartesian coordinate system will be:

$$\overrightarrow{a} \cdot \overrightarrow{b} = (a_x b_x + a_y b_y) = a_x b_x \qquad (2.23)$$

but the component b_x is the *projection* of the vector $\overrightarrow{b}$ onto the x axis. Therefore, if α is the angle between the two vectors, we have that $b_x = \|\overrightarrow{b}\| \cos \alpha$. We already know that $\|\overrightarrow{a}\| = a_x$ and so the scalar product is:

---

[8]In reality we will see later that it is possible to generate a *pseudoscalar* if we make the scalar product of a vector and a *pseudovector* (defined later).

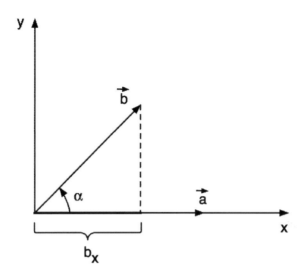

FIGURE 2.5 Geometric interpretation of the dot product as projection.

$$\vec{a} \cdot \vec{b} = a_x b_x = \|\vec{a}\| \|\vec{b}\| \cos\alpha \qquad (2.24)$$

Eq. 2.24 tells us that the scalar product of two orthogonal (perpendicular) vectors is zero.

In summary, we have that the dot product of two vectors can be calculated in two different equivalent ways:

$$\vec{a} \cdot \vec{b} = (a_x b_x + a_y b_y)$$
$$\vec{a} \cdot \vec{b} = \|\vec{a}\| \|\vec{b}\| \cos\alpha \qquad (2.25)$$

The first by using the coordinates of the vector in a given coordinate system, the second by using the magnitudes and angle between the two vectors (once their bases are put together).

The other operation involving two vectors is the so-called *vector product* . The vector product can be defined by considering two vectors $\vec{a}$ and $\vec{b}$ not having the same direction. In this case, these two vectors uniquely identify a plane. We want to build a vector $\vec{c}$ that is perpendicular to this plane, i.e. we request that $\vec{c}$ is both perpendicular to $\vec{a}$ and $\vec{b}$. This problem can be solved algebraically by requesting that the scalar products $\vec{a} \cdot \vec{c} = \vec{b} \cdot \vec{c} = 0$.

Using eq. 2.21 (because now we have vectors spanning the 3-dimensional space), we have:

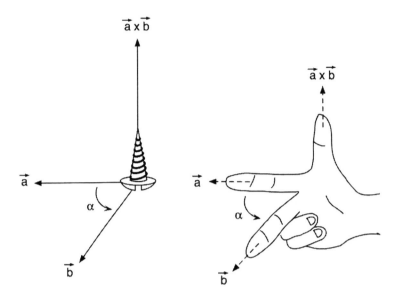

FIGURE 2.6 Right-hand rule in the vector product.

$$Xa_x + Ya_y + Za_z = 0$$
$$Yb_x + Yb_y + Zb_z = 0$$

(2.26)

where for clarity we called the components of the vector $\vec{c}$, (X, Y, Z). The system of equations 2.26 contains two equations with three unknowns. This means that there are infinite solutions (coordinates) that represent the vector perpendicular. All the solutions are parallel to each other and therefore we should not be surprised. To find a convenient solution, let us multiply the first equation of the system 2.26 by b_x and the second equation by a_x:

$$Xa_x b_x + Ya_y b_x + Za_z b_x = 0$$
$$Ya_x b_x + Ya_x b_y + Za_x b_z = 0$$

(2.27)

If we subtract the second equation from the first and after a little bit of algebra we find two solutions:

$$X = \frac{(a_x b_z - a_z b_y)}{(a_x b_y - a_y b_x)} Z$$
$$Y = -\frac{(a_x b_z - a_z b_x)}{(a_x b_y - a_y b_x)} Z$$

(2.28)

If we choose the particular solution $Z = (a_x b_y - a_y b_x)$, then the vector perpendicular has coordinates:

$$X = (a_y b_z - a_z b_y)$$
$$Y = -(a_x b_z - a_z b_x) \qquad (2.29)$$
$$Z = (a_x b_y - a_y b_x)$$

When we made the arbitrary choice $Z = (a_x b_y - a_y b_x)$ we also defined the orientation of the vector resulting from the vector product. In fig. 2.6 we see the geometry of the vector product. The two vectors $\vec{a}$ and $\vec{b}$ identify a plane. The vector $\vec{a} \times \vec{b}$, constructed to be perpendicular to the plane, can point either upwards or downwards. The choice we made is to have it point upwards. With this choice, the three vectors identify a *right-handed coordinate system*: to decide where the vector product is pointing, we align the first vector in the product ($\vec{a}$) with the index finger of a **right hand** (right panel in fig. 2.6). The second vector in the product ($\vec{b}$) is then aligned along the middle finger. With this convention, the thumb indicates the direction of the vector product. It follows immediately that the vector product satisfies:

$$\vec{a} \times \vec{b} = -\vec{b} \times \vec{a} \qquad (2.30)$$

i.e. is not *commutative*. You can convince yourself by now aligning the vector $\vec{b}$ with the index finger and the vector $\vec{a}$ with the middle finger. Your thumb now will point downwards instead of upwards. An alternative way to avoid overexerting your wrists can be seen in the left panel of fig. 2.6: a right-handed screw turned anti-clockwise (from $\vec{a}$ to $\vec{b}$) will advance upwards. If you exchange the two vectors the screw will advance downwards.

We now show that, in analogy with eq. 2.25, there are two equivalent ways to calculate the vector product. Eq. 2.29 already tells us how to calculate the coordinates of the vector product given the coordinates of the two vectors being multiplied by each other. We now show that there is a way to calculate the magnitude of the vector product given the magnitude and direction of the two vectors being multiplied. The following derivation is quite cumbersome and lengthy and that's probably why it is not often reported in textbooks.

To make the expressions a bit easier to follow we now consider two vectors $\vec{a}$ and $\vec{b}$ with coordinates respectively (a_1, a_2, a_3) and (b_1, b_2, b_3). From eq. 2.29 let's write the square of the magnitude of the vector product:

$$\begin{aligned}
\|\vec{a} \times \vec{b}\|^2 &= (a_2 b_3 - a_3 b_2)^2 + (a_3 b_1 - a_1 b_3)^2 + (a_1 b_2 - a_2 b_1)^2 \\
&= a_2^2 b_3^2 - 2 a_2 a_3 b_2 b_3 + a_3^2 b_2^2 \\
&\quad + a_3^2 b_1^2 - 2 a_1 a_3 b_1 b_3 + a_1^2 b_3^2 \\
&\quad + a_1{}^2 b_2^2 - 2 a_1 a_2 b_1 b_2 + a_2^2 b_1^2 \\
&= a_1^2 (b_2^2 + b_3^2) + a_2^2 (b_1^2 + b_3^2) + a_3^2 (b_1^2 + b_2^2) \\
&\quad - 2(a_2 a_3 b_2 b_3 + a_1 a_3 b_1 b_3 + a_1 a_2 b_1 b_2)
\end{aligned} \tag{2.31}$$

Let us now write the square of the magnitude of the dot product:

$$\begin{aligned}
(\|\vec{a}\|\|\vec{b}\| \cos \alpha)^2 &= (a_1 b_1 + a_2 b_2 + a_3 b_3) \cdot (a_1 b_1 + a_2 b_2 + a_3 b_3) \\
&= a_1^2 b_1^2 + a_1 a_2 b_1 b_2 + a_1 a_3 b_1 b_3 \\
&\quad + a_2^2 b_2^2 + a_1 a_2 b_1 b_2 + a_2 a_3 b_2 b_3 \\
&\quad + a_3^2 b_3^2 + a_1 a_3 b_1 b_3 + a_2 a_3 b_2 b_3 \\
&= a_1^2 b_1^2 + a_2^2 b_2^2 + a_3^2 b_3^2 \\
&\quad + 2(a_1 a_2 b_1 b_2 + a_1 a_3 b_1 b_3 + a_2 a_3 b_2 b_3)
\end{aligned} \tag{2.32}$$

Now we add the two equations 2.31 and 2.32:

$$\begin{aligned}
\|\vec{a} \times \vec{b}\|^2 + \|\vec{a}\|^2 \|\vec{b}\|^2 \cos^2 \alpha &= a_1^2 (b_1^2 + b_2^2 + b_3^2) \\
&\quad + a_2^2 (b_1^2 + b_2^2 + b_3^2) \\
&\quad + a_3^2 (b_1^2 + b_2^2 + b_3^2) \\
&= \|\vec{a}\|^2 \|\vec{b}\|^2
\end{aligned} \tag{2.33}$$

Eq. 2.33 reduces to:

$$\begin{aligned}
\|\vec{a} \times \vec{b}\|^2 + \|\vec{a}\|^2 \|\vec{b}\|^2 \cos^2 \alpha &= \|\vec{a}\|^2 \|\vec{b}\|^2 \\
\|\vec{a} \times \vec{b}\|^2 &= \|\vec{a}\|^2 \|\vec{b}\|^2 (1 - \cos^2 \alpha) \\
\|\vec{a} \times \vec{b}\|^2 &= \|\vec{a}\|^2 \|\vec{b}\|^2 \sin^2 \alpha
\end{aligned} \tag{2.34}$$

Taking the square root of the last equation, we have:

$$\|\vec{a} \times \vec{b}\| = \|\vec{a}\|\|\vec{b}\| \sin \alpha \tag{2.35}$$

which is the analog of the second equation in eq. 2.25 for the vector product.

There is a geometric interpretation of the vector product (see fig. 2.7). The vector product is a vector whose magnitude is the area of the parallelogram identified by the two vectors $\vec{u}$ and $\vec{v}$. We see immediately that the area of

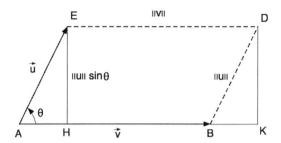

FIGURE 2.7 Geometric interpretation of the vector product .

the parallelogram $ABDE$ is equal to the area of the rectangle $HKDE$ whose area is indeed $\|\vec{u}\| \, \|\vec{v}\| \, \sin\theta$. The direction is given by the right hand rule.

Let's conclude this section by stating a useful formula for the vector product of three vectors $\vec{a}$, $\vec{b}$ and $\vec{c}$:

$$\vec{a} \times (\vec{b} \times \vec{c}) = \vec{b}(\vec{a} \cdot \vec{c}) - \vec{c}(\vec{a} \cdot \vec{b}) \tag{2.36}$$

2.1.3   Differentials and Derivatives of Vectors

Now that we have defined vectors and some operations, we need to see what we can do with them. If we want to study the motion of planets, for example, we need to know how things change with time. The planet's position, for example, can be indicated with three numbers corresponding to its coordinates in an x, y, z Cartesian coordinate system. We understand the motion if we know how to calculate the position of the planet after a certain interval of time. We have seen that we can associate a vector to displacements and so if we want to know how the position of the planet changes with time we need to study how the position vector changes with time, i.e. we need to understand the derivative of a vector quantity.

Let us first introduce the concept of a *differential*[9]. Let $f(x)$ represent a function that can be differentiated on an open interval containing the variable x[10]. We *define* the differential dx as *any* non-zero real number[11]. We *define* the differential of the function $f(x)$ as:

$$dy = f'(x)dx = \frac{df}{dx}dx \tag{2.37}$$

Notice that the differential of the independent variable x is just a number

---

[9]The definition we give is not rigorous but good enough for the mathematical functions found in the physical problems that we will encounter in celestial mechanics.

[10]An open interval is an interval that does not contain its endpoints. In our case it is enough to say that the function $f(x)$ is differentiable between $-\infty$ and $+\infty$.

[11]We always consider this number very small and tending to zero.

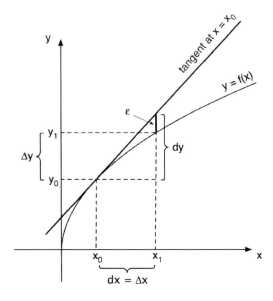

FIGURE 2.8 Geometrical interpretation of the differential $dy = f'(x)dx$ of a function $f(x)$.

while the differential of a dependent variable, i.e. a variable that is a function of one (or more) variables, is defined by eq. 2.37. In fig. 2.8 we can see a geometric interpretation of the concept of the differential. The function $f(x)$ goes smoothly (differentiable) from the point (x_0, y_0) to (x_1, y_1). Suppose we want to estimate the value of the function $f(x_1)$ evaluated at x_1, knowing its value $f(x_0)$ at x_0. From the figure we see that a good approximation would be to calculate the tangent to $f(x)$ so as to estimate $f(x_1)$ with the value of the tangent at x_1. This estimate produces an error equal to ϵ. In fact we have:

$$dy = (y_1 - y_0) - \epsilon$$
$$= \Delta y - \epsilon \approx \Delta y \qquad (2.38)$$

where ϵ goes smaller and smaller as x_1 approaches x_0. The approximation would be $dy \approx \Delta y$ and becomes better and better as $x_1 \to x_0$ with $\epsilon \to 0$. If we accept an approximate value for the function $f(x)$ at x_1 then we can "control" the accuracy by calculating the error ϵ. If the error is acceptable, then we can evaluate the function $f(x)$ at x_1 by simply calculating the value at x_1 along the tangent instead of the function $f(x)$. The equation of the tangent at x_0 of the function $f(x)$ is given by:

$$(y_1 - y_0) = m(x_1 - x_0)$$

$$m = \frac{dy}{dx} \tag{2.39}$$

$$dy = \frac{dy}{dx} dx = f'(x) dx$$

The described procedure to approximate the value of a function with the value calculated along the tangent is called *linear approximation*.

We now give a few useful relations involving differentials . Given two functions u and v and two constants p and q, it is easy to show that differentials obey the following rules:

$$d(p) = d(q) = 0$$
$$d(u + v) = du + dv$$
$$d(qu) = qdu \tag{2.40}$$
$$d(uv) = vdu + udv$$
$$d(u/v) = \frac{(vdu - udv)}{v^2}$$

Let us now go back to the meaning of the derivative of a vector. Suppose we have a vector $\vec{u}$ that changes with time. We can write that $\vec{u} = \vec{u}(t)$. What is the derivative of this vector? Extending eq. 1.27 to the vector $\vec{u}$ we can write:

$$\frac{d\vec{u}}{dt} = \lim_{\Delta t \to 0} \frac{\vec{u}(t + \Delta t) - \vec{u}(t)}{\Delta t} \tag{2.41}$$

If we think of the vector $\vec{u}$ in terms of magnitude and direction, we can decompose its variation with time separately for the magnitude and for the direction. We can then study the two variations independently. In fig. 2.9, left panel, we see how a change of magnitude of a vector $\vec{u}$ produces a vector $d\vec{u}$ which is *parallel* to the vector $\vec{u}$. The right panel shows how a change in direction produces a vector $d\vec{u}$ which is now *perpendicular* to the vector $d\vec{u}$ when the angle $d\alpha$ gets smaller and smaller. Now suppose that the vector $d\vec{u}$ is rotating counterclockwise without changing its magnitude. It is obvious that the tip of the vector will describe a circumference. If the rate of change of the angle α is constant, then the tip of the vector is describing what is commonly called a *uniform circular motion*, i.e. $\|\vec{u}\|$=constant and $\frac{d\alpha}{dt}$=constant.

Let's study mathematically (the physics will come later) a bit more in detail the trajectory of a point in a 2-dimensional plane. The motion of a point is fully described if we give the $x = x(t)$ and $y = y(t)$ coordinates as a function of time (see fig. 2.10).

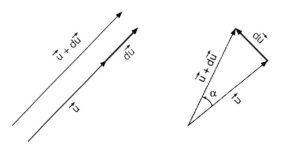

FIGURE 2.9 Variation of a vector. Left panel shows the rate of change of the vector with change of magnitude while the right panel shows the rate of change of the vector with change of direction.

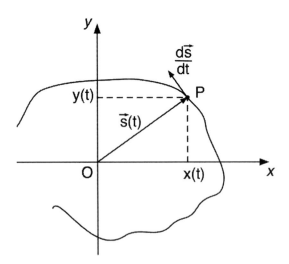

FIGURE 2.10 Trajectory of a moving point P on a x, y coordinate system. The coordinates of the point P, during its motion are both function of time, $x = x(t), y = y(t)$.

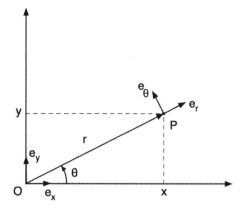

FIGURE 2.11 Cartesian and Polar coordinate systems in two dimensions. Note the orientation of the unit vectors.

We have already seen how to represent symbolically a vector by writing:

$$\vec{s}(t) = \begin{pmatrix} x(t) \\ y(t) \end{pmatrix} \tag{2.42}$$

The vector $\vec{s}$ changes with time during the motion of the point P and its coordinates $x = x(t)$ and $y = y(t)$ must change with time accordingly. Let's now write down the derivative of the vector expressed in 2.42:

$$\frac{d}{dt}\vec{s}(t) = \frac{d}{dt}\begin{pmatrix} x(t) \\ y(t) \end{pmatrix} = \begin{pmatrix} \frac{dx}{dt} \\ \frac{dy}{dt} \end{pmatrix} \tag{2.43}$$

Eq. 2.43 shows that the derivative of a vector $\vec{s}$ can be represented by a new vector whose components are the derivative of the components of $\vec{s}$. This new vector is tangent to the curve in the point P and tells us the instantaneous velocity (magnitude and direction) of the point P. We know already that the scalar product $\vec{s} \cdot \frac{d\vec{s}}{dt} = 0$, meaning that the two vectors are perpendicular.

2.1.4 Polar and Cylindrical Coordinates

We will see that in the study of the orbital motion of planets there are a set of equations that need to be solved to obtain all the orbital parameter. We have used, so far, Cartesian coordinates to describe various mathematical quantities. However, quite often, equations written in Cartesian coordinates can become quite difficult to treat. It is possible to express equations and various important vectors in a different coordinate system which is more natural when treating orbital motion: polar coordinates .

With reference to fig. 2.11, we see that the position of a point P is uniquely

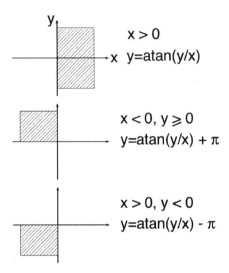

FIGURE 2.12 The function y=atan2(x,y). From top to bottom, when $x > 0$ and y is positive or negative, the atan2 is just the atan. When $x < 0$ and $y \geq 0$, then we need to add π to the calculated value of the atan. Finally, when $x > 0$ and $y < 0$, then we need to subtract π to the calculated value of atan.

identified by either giving its Cartesian coordinates (x, y), or its polar coordinate (r, θ). In the polar coordinate system, the point P is identified by giving the distance between a reference point O and the point P and the angle θ between a reference axis and the polar axis. Simple trigonometry allows us to write down the equations of transformation between Cartesian and polar coordinate systems. Knowing the polar coordinate we can find the Cartesian by using:

$$x = r \cos \theta$$
$$y = r \sin \theta$$
(2.44)

The reverse transformation, i.e. Cartesian-to-polar, is a bit more complicated because of an ambiguity in the angle θ. The Cartesian to polar transformations are given by:

$$r = \sqrt{x^2 + y^2}$$
$$\theta = \text{atan2}(y, x)$$
(2.45)

where the function atan2(y, x) is defined in fig. 2.12.

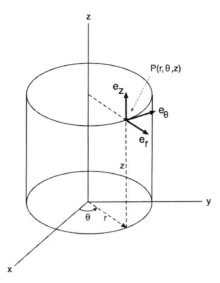

FIGURE 2.13 A point P in cylindrical coordinates.

The extension to three dimensions of polar coordinates is called **cylindrical coordinates**. A point P in space is identified with three coordinates (r, θ, z) as represented in fig. 2.13. The coordinate transformation formulas between Cartesian and cylindrical coordinate are:

$$
\begin{aligned}
r &= \sqrt{x^2 + y^2} \\
\theta &= \operatorname{atan2}(y, x) \\
z &= z
\end{aligned}
\tag{2.46}
$$

The unit vectors of cylindrical coordinates are $\vec{e_r}, \vec{e_\theta}, \vec{e_z}$ and they form a right-handed coordinate base for which $\vec{e_r} = \vec{e_\theta} \times \vec{e_z}$. Using the definition of $\sin \theta$ and $\cos \theta$ we have:

$$
\begin{aligned}
\vec{e_r} &= \vec{e_x} \cos \theta + \vec{e_y} \sin \theta \\
\vec{e_\theta} &= -\vec{e_x} \sin \theta + \vec{e_y} \cos \theta \\
\vec{e_z} &= \vec{e_z}
\end{aligned}
\tag{2.47}
$$

and using the chain rule (eq. 1.30), we can calculate the time derivatives of the unit vectors:

$$
\begin{aligned}
\frac{d\vec{e_r}}{dt} &= \frac{d\vec{e_r}}{d\theta} \frac{d\theta}{dt} = \vec{e_\theta} \frac{d\theta}{dt} \\
\frac{d\vec{e_\theta}}{dt} &= \frac{d\vec{e_\theta}}{d\theta} \frac{d\theta}{dt} = -\vec{e_r} \frac{d\theta}{dt}
\end{aligned}
\tag{2.48}
$$

2.1.5   Vectors in Physics

Why are vectors important in physics and in particular in the study of the motion of planets in the solar system? Because many physical quantities involved in the study of motion cannot be described by just a numerical quantity, but need more information to be completely determined. In particular, vectors are quantities that are fully described if we give the magnitude and the direction. An arrow is a good visual representation of a vector because the length can be made proportional to the magnitude and the tip of the arrow indicates the orientation.

There is one more important property of vectors. All the equations needed to define the motion of celestial objects in our solar system consist of a relationship among various vectors. This means that if we understand how to manipulate vectors algebraically then we can manipulate the vector equation describing the dynamics of our solar system objects.

Finally, if we express physical laws as relationships between vectors, these relationships are **independent** from the specific coordinate axes that we choose to represent the vectors. This is very powerful because we can choose the most convenient coordinate system depending on the specific physical system and its properties.

Notice that not all the laws in physics can be expressed in terms of vectors. *Tensors* are instead used and they are more general entities for which the vector is a particular case. [12]

2.1.6   Polar and Axial Vectors

When we talked about operations with vectors we introduced two ways to multiply two vectors: scalar product and vector product . These two operations, although both containing the word "multiply", are inherently different. The scalar product of two vectors, for example, generates a number or scalar. The vector product of two vectors generates another vector. We will see that vectors obtained through vector products are different from vectors like displacement, velocity or acceleration. We have studied how the components of a vector change under a coordinate transformation like a simple translation or a rotation. Let's look at the physics for a moment. Let us assume that the distant stars are so distant that they do not move appreciably and therefore they constitute a reference system to which we refer the motions on Earth[13].

Let us study what happens to vectors once subject to a special kind of coordinate transformation: **parity**. A parity transformation is a change of

---

[12]In the elementary discussion of this book we will not consider tensor algebra and its application in physics. See further readings at the end of this chapter.

[13]For small Earth's motion this is quite a good assumption. However, if we have to consider, for example, the motion of the stars in our galaxy, then the distant stars reference frame would not be adequate anymore and we should change to a system composed of all distant galaxies. This system might also not be adequate if we study the large-scale structure of the Universe.

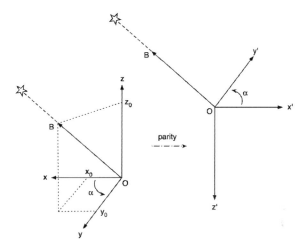

FIGURE 2.14 Parity transformation, $x \to x' = -x$, $y \to y' = -y$ and $z \to z' = -z$. Notice that the coordinate system in the left panel is right-handed while the one in the right panel is left-handed. A polar vector is represented.

coordinates obtained by reversing the directions of all the coordinate axes. If we indicate with P the action of reversing the directions of the coordinate axes, we have

$$\begin{pmatrix} x' \\ y' \\ z' \end{pmatrix} = P \begin{pmatrix} x \\ y \\ z \end{pmatrix} = \begin{pmatrix} -x \\ -y \\ -z \end{pmatrix} \qquad (2.49)$$

A vector-like $\vec{u}$ will change its coordinates according to eq. 2.49 such that:

$$\vec{u'} = P\vec{u} = -\vec{u} \qquad (2.50)$$

Eq. 2.50 tells us that in order for the vector $\vec{u}$ to keep the original magnitude and direction, *in the new coordinate system* where the axes are flipped, the coordinates of $\vec{u}$ need to be flipped as well.

Now let us consider a simple displacement vector describing a space ship traveling from the point O (origin of the coordinate system) to the point B in space towards the star Alpha Centauri (see fig. 2.14). The displacement vector $\overrightarrow{OB}$ with coordinates (x_0, y_0, z_0) indicates that in a certain time interval the spaceship has traveled from the point A to the point B in space. Obviously, such displacement cannot depend on the choice of coordinate system and therefore, in order for the vector $\overrightarrow{OB}$ to still point in the same direction and with the same magnitude, its coordinates have to change from (x_0, y_0, z_0) to

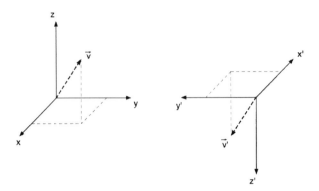

FIGURE 2.15 An axial vector under parity transformation.

$(-x_0, -y_0, -z_0)$. Same behavior must be expected for other vectors like the velocity or the acceleration towards Alpha Centauri. Vectors that change their coordinates according to eq. 2.49 are called **polar vectors** .

There are other vectors[14] that, under parity transformation, do not change sign but flip in the same way as the coordinates. This is the case of vectors generated by the vector product of two polar vectors $\vec{u}$ and $\vec{v}$:

$$P(\vec{u} \times \vec{v}) = (\vec{u'} \times \vec{v'}) = (-\vec{u} \times -\vec{v}) = (\vec{u} \times \vec{v}) \tag{2.51}$$

Vectors transforming their coordinates according to eq. 2.51 (see fig. 2.15) are called **axial vectors** [15]. We will see that polar vectors are mostly associated with forces that generate motion along the same direction as the vector. For example, a simple impulse on a billiard ball coming from the cue stick will result in the ball moving along the direction of the motion of the cue (if no spinning effect is imparted to the ball). Example of polar vectors are: displacement, velocity, acceleration, and gravitational attraction.

The vector multiplication of two polar vectors, as in eq. 2.29, generates an axial vector. It can be shown that we must have:

1. (polar vector) × (polar vector) = (axial vector)

2. (axial vector) × (axial vector) = (axial vector)

3. (axial vector) × (polar vector) = (polar vector)

4. (polar vector) × (axial vector) = (polar vector)

---

[14] For the mathematically inclined, the objects we will be talking about now are not proper vectors. It happens that in a 3-dimensional space they behave very much like vectors, while in reality they are antisymmetric tensors.

[15] Also called pseudovectors.

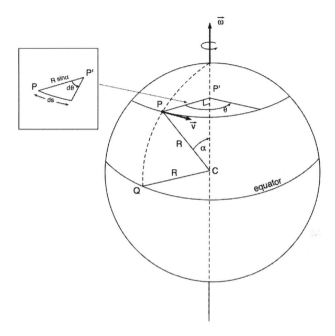

FIGURE 2.16 Angular velocity due to the rotation of the Earth.

Axial vectors are associated mostly with rotations. Let us try to define an angular motion in the 3-dimensional space with a vector. Let us consider the rotation of the Earth and a point P on its surface located in the Northern Hemisphere at an angle α from the North Pole[16]. Due to the rotation of the Earth, this point will rotate at an angle θ during a certain period of time. We know already that the point P will return to its original position (relative to the fixed stars, for example) after about 24 hours. We can calculate the amount of space that the point P will travel along the arc length after a rotation of an angle θ. If the point Q is on the equator, as in the fig. 2.16, it will travel the maximum distance $2\pi \cdot \overline{CQ} = 2\pi R \approx 40,000$ km, where R is the radius of the Earth. This means that the point Q will travel the maximum circumference of the Earth. However, if the point P is at a certain angle α from the North Pole, then it will travel on a circumference that has a radius equal to $\overline{PP'} = R \sin \alpha$. Let us now calculate a small arc length ds (see square insert in the upper left corner of fig. 2.16).

$$ds = d\theta \, R \sin \alpha \qquad (2.52)$$

We can divide the above equation by the differential dt:

---

[16]This angle is 90° minus the latitude of the place.

$$\frac{ds}{dt} = \frac{d\theta}{dt}\, R\sin\alpha = \omega\, R\sin\alpha \tag{2.53}$$

The quantity $\frac{ds}{dt}$ is the instantaneous velocity $\vec{v}$ tangent to the point P. Eq. 2.53 has the same form as eq. 2.35 defining the vector product, if we identify ω and R with the magnitude of two vectors, respectively $\vec{\omega}$ and $\vec{R}$. The vector $\vec{R}$ is clearly the radius R expressed as a polar vector. We therefore re-write eq. 2.53 as:

$$\vec{v} = \vec{\omega} \times \vec{R} \tag{2.54}$$

What is the vector $\vec{\omega}$? First, we notice that $\vec{\omega}$ is perpendicular to both $\vec{R}$ and $\vec{v}$, i.e. is perpendicular to the plane where the rotation is happening. In the vector product in eq. 2.54, the output vector of the vector product is a polar vector. Since $\vec{R}$ is also a polar vector, then $\vec{\omega}$ must be an axial vector because of rule 4 above. $\vec{\omega}$ is aligned with the axis of rotation of the Earth and its magnitude gives the radians per second of rotation. So, $\vec{\omega}$ is called **angular velocity** , and it is a vector aligned with the axis of rotation.

2.2 NEWTON'S LAWS AND GRAVITY

In the previous section we have developed a few useful mathematical tools. We now switch to physics and study a few concepts that will be essential to understand the motion of celestial bodies in the next chapter. It is a common experience among physicists that the more basic the quantity we try to define the more difficult the task is. In this section we state the Newton's three conservation laws, mostly used in Newtonian mechanics (energy, momentum, angular momentum), and Newton's gravitation law. Armed with these formidable tools we will be ready to finally study in detail the motion of planets under the influence of gravity.

Sky objects like stars and planets appear to move. They appear to rotate around the north celestial pole (NCP)[17]. This motion is clearly a geometric effect due to the rotation of the Earth. A more interesting question is: do the stars move or are they fixed? If they move, what kind of motion is involved and what is the origin of such motion?

We already know that planets move and they move into a closed path around the Sun. If the Sun was not present, the planets would continue to move on a straight line forever. In order to understand the motion, we need the laws of motion, i.e. statements and equations that will help us calculate how objects move. There are three laws and they were first presented in the year 1687 by Isaac Newton [9].

The first law of motion was actually first formulated by Galileo Galilei as his inertia definition. It can be stated as follows:

---

[17]The North Celestial Pole is the point in the sky obtained by extending the Earth's rotation axis into the sky where it intersects the celestial sphere.

FIGURE 2.17 Sir Isaac Newton (1643 - 1727) is undoubtedly one of the greatest scientists of all time. He laid the foundations of classical mechanics and invented differential calculus, together with Gottfried Wilhelm Leibniz. He has worked in optics, fluid dynamics and theory of thermal conductivity.

FIGURE 2.18 Galileo Galilei (1564 - 1642) is recognized as the father of observational astronomy. He was the first to build himself a telescope and use it to study the sky objects. Galileo was a strong advocate of heliocentrism which caused him to be subject to the Roman Inquisition in 1615. He was forced to recant his views and spent the rest of his life under house arrest.

First Law: *Every object will remain at rest or in uniform motion in a straight line unless compelled to change its state by the action of a net external force.*

We can use the first law to define the concept of *force*: when a body changes its state of motion, i.e. accelerates, decelerates, change direction or a combination of the above, then a force is applied. There are contact forces when the object has changed its state of motion as a result of having another body coming in direct contact. Think, for example, of the change of trajectory of a snooker ball when is hit by another snooker ball. There are also non-contact forces able to act upon a body apparently without contact and across empty space, like gravity or magnetic/electric forces.

What kind of quantity is a force? If we kick a football laying on the grass, we know that the heavier the ball, the shorter the distance it will travel before stopping. So the force must be proportional to the mass of the ball. We also know that if we can kick the ball in a certain direction, the ball will go along exactly that direction. So the force has a magnitude and a direction: it is a vector. This means that to define a force, we need to tell not only how strong it is but also in which direction it is acting.

We now introduce an important concept: **momentum** of a particle of mass m. The momentum of a particle of mass m moving with a velocity $\vec{v}$ is a vector (polar) defined by:

$$\vec{p} = m \cdot \vec{v} \qquad (2.55)$$

Newton stated the second law of motion as a relationship between force and momentum:

Second Law: *The change of motion, i.e. momentum of an object, is proportional to the force impressed upon it, and is made in the direction of the straight line in which the force is impressed.*

The second law can be expressed mathematically as:

$$\vec{F} = \frac{d\vec{p}}{dt} \qquad (2.56)$$

where $\vec{F}$ is the force and $\vec{p}$ is the *momentum*, defined as the product of the *mass* times the *velocity*. A more familiar form of eq. 2.56 is shown in eq. 2.57 if we assume that the mass m is constant with time:

$$\vec{F} = \frac{d\overrightarrow{(mv)}}{dt} = m\frac{d\vec{v}}{dt} = m\vec{a} \qquad (2.57)$$

Eq. 2.57 tells us that the force is a vector; it accelerates the body to which it is applied and is aligned along the direction of the acceleration. In the SI system [18] the unit of force is found by looking again at its definition: it is the force that when applied to a body of mass 1 kg, impresses an acceleration of 1 meter per sec^2. It should not surprise us that such a unit is called 1 Newton to celebrate Isaac Newton who first proposed his second law of motion.

For completeness we enunciate also the third law of motion:

Third Law: *To every action there is always opposed an equal reaction. Alternatively the mutual actions of two bodies on each other are always equal and opposite.*

How do we use Newton's laws to solve dynamical problems, i.e. how do we use equation 2.56? This equation tells us that if we know the force vector $\vec{F}$ and the mass m of the body, then we can calculate the acceleration. Let us study the simple situation for which there are no *external* forces, i.e. $\vec{F} = 0$. For simplicity, let us restrict the system to just the x coordinate, i.e. the motion is only along the x axis. Eq. 2.56 or 2.57 become:

---

[18]The SI is the International System of units where we measure physical quantities in terms of seven base units: ampere (electrical current), kelvin (temperature), second (time), meter (length), kilogram (mass), candela (luminous intensity) and mole (amount of substance).

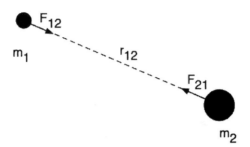

FIGURE 2.19 Geometry of the gravitational forces acting between two
bodies of masses m_1 and m_2.

$$m\frac{d^2x}{dt^2} = 0$$
$$m\frac{dx}{dt} = const$$

(2.58)

So the body of mass m in absence of external forces maintains a constant
speed. This is Newton's first law, also known as Galileo's inertia definition.
Another simple example concerns the weight force $\overrightarrow{P} = m\overrightarrow{g}$ which is the
force due to the Earth's gravitational attraction at the surface of the Earth.
We will talk extensively of gravitational attraction in the rest of this book.
So we now have Newton's second law and a force specified. We can therefore
write:

$$m\frac{d^2x}{dt^2} = mg$$

(2.59)

where g is the gravitational acceleration at the surface of Earth equal to
$g = 9.81 \; meter \cdot s^{-2}$. The mass m cancels out telling us that the free fall of
bodies of different masses follows exactly the same trajectory as notoriously
shown by Galileo and, more recently, by the Apollo 15 crew on the surface of
the Moon with a hammer and a feather. Eq. 2.59 describes the motion of a
body on the surface of the Earth subject only to the attraction of the Earth
and neglecting air resistance and Earth's rotation. Eq. 2.59 is an example of a
differential equation where the unknown is the function $x = x(t)$ describing

the position x of the free falling body. It is straightforward to check that the solution to eq. 2.59 is:

$$x(t) = x_0 + v_0 t + \frac{1}{2}gt^2 \qquad (2.60)$$

where the coordinate x now represents the vertical coordinate towards the center of the Earth and x_0 and v_0 are respectively the initial position and initial velocity of the free falling body.

Eq. 2.56 does not tell us the effect of a non-contact force between two bodies of masses m_1 and m_2. In order to understand the motion of celestial bodies, for example, we need to specify what force exists between the Sun and a planet or the Earth and the Moon, etc. Newton, in his *Principia* book, first describes such a force (gravity) acting between two bodies of masses m_1 and m_2, separated by a distance r_{12}:

$$\overrightarrow{F}_{12} = G\frac{m_1 m_2}{r_{12}^2}\overrightarrow{k}_{12} \qquad (2.61)$$

where $G = 6.674 \cdot 10^{-11} \ N \cdot kg^{-2} \cdot m^2$ and $\overrightarrow{k_{12}}$ is the unit vector along the direction connecting the two masses. $\overrightarrow{F}$ is a polar vector in the sense that the force is directed along the line 12. Notice the relatively small value of the constant G, which is balanced by the large mass of the Earth, for example, to justify the fact that we "feel" gravity anyway. In conclusion, we have seen that Newton's laws provide us with differential equations once we have understood all the forces at play, i.e. we have defined all the vectors. The solutions to these differential equations represent the motion of the bodies. Having defined the force due to gravity between masses, we are now in a position to study the orbital motion of planets.

2.3   THE CONCEPT OF MASS

In the previous section we have introduced the concept of momentum of a particle as the product of a scalar quantity, its mass, times a quantity, its velocity. We already know that multiplying a vector by a scalar produces another vector and therefore the momentum is a vector. We now need to define the mass of a body. Newton's second law tells us that if we apply a force $\overrightarrow{F}$ to a body at rest, it will start to accelerate with acceleration $\overrightarrow{a}$. We can define the mass m as follows:

$$m = \frac{\|\overrightarrow{F}\|}{\|\overrightarrow{a}\|} \qquad (2.62)$$

Eq. 2.62 tells us that, given the same force $\overrightarrow{F}$ applied to different bodies, we will measure smaller and smaller accelerations for larger and larger masses. The mass measured in this way is a measure of the **inertia** of a body, i.e. its resistance to be accelerated, and it is called **inertial mass** . It is quite evident that it is easier to accelerate a motorcycle rather than a car simply because the motorcycle is lighter than the car.

There is another kind of force that bodies are subject to: the weight. We are all familiar with the concept of weight by simply comparing the amount of effort we have to produce to hold different balls in our hands. A ball made of iron will require much more effort than a ball of wool. The force that the balls are subject to is called **gravitational force** and is directed towards the center of the Earth. We might be tempted to use Newton's second law with this force to define the masses of the different balls. However, we know that all the bodies subject to this force have all the same acceleration! In fact, the accelerations of bodies left to fall to the ground are all the same no matter how big or small or dense the body is. So what is going on?

First, let us define the weight $\overrightarrow{P}$ of an object[19] using eq. 2.57:

$$\overrightarrow{P} = m_g \overrightarrow{g} \qquad (2.63)$$

where g is the *gravitational acceleration* on the surface of the Earth. The force $\overrightarrow{P}$ is a *non-contact* force and it is due to the gravitational attraction of the Earth.

The only way to make sure that the acceleration $\overrightarrow{g}$ in eq. 2.63 is constant is by assuming that the weight force $\overrightarrow{P}$ is proportional to the mass m_g. Notice that the gravitational mass defined in this way is different from the inertial mass defined earlier. In fact, this mass is called **gravitational mass** [20]. Are gravitational mass and inertial mass of a body different? The answer is no: the equivalence of gravitational mass and inertial mass is a well-established experimental fact: as of 2018 [10] the two masses are measured to be equal[21] to better than 1 part in 10^{13}. Albert Einstein started from this equivalence to build his general relativity theory.

Think About It...

We have stated that axial vectors are somehow related to rotations. The magnetic field can be considered as an axial vector and, interestingly, can be generated by having electrons rotating in a conducting loop.

FURTHER READING

Goldstein, H. (2002), *Classical Mechanics*. Addison-Wesley.

---

[19] Remember that the weight is a force and therefore is a vector quantity.
[20] More precisely, this is the definition of Passive Gravitational Mass.
[21] This is also-called WEP for *weak equivalence principle*.

Joag, P.S. (2016), *An Introduction to Vectors, Vector Operators and Vector Analysis.* Cambridge University Press.

Rupert Hall, A. (1981). *From Galileo to Newton.* Dover Publications Inc.

Taylor, J. (2005). *Classical Mechanics.* University Science Books.

Celestial Mechanics

CONTENTS

THE motion of planets is regulated by Kepler's laws. Kepler was the assistant of the Danish astronomer Tycho Brahe who spent several years making accurate measurements of the position of stars and planets. Kepler was a good mathematician and studied Tycho's data with great attention. He found that the motion of planets obeyed very well three *empirical* laws. Kepler's laws provided the foundation for Newton's theory of gravitation.

We will see that Kepler's laws can be obtained from Newton's laws, and not vice versa, which means that Newton's laws are more fundamental than Kepler's laws. Most of the dynamical problems are usually treated by assuming Newton's laws plus a few extra laws called *conservation laws*: momentum, angular momentum and energy. The usage of conservation laws and Newton's laws is the way in which orbital motion of planets is studied in undergraduate courses.

Then, when progressing to more modern mathematics and physics, it is found that Newton's laws are not the most fundamental but can be obtained by a more general principle: the principle of least action[1]. This principle provides a beautiful description of dynamical phenomena that contains not only Newton's laws but also the conservation laws. In this sense, the principle of least action is more fundamental than Newton's laws. This principle is so general that it actually has a fundamental role not only in classical mechanics, but provides a direct link between classical mechanics and quantum physics.

In this chapter we will introduce immediately the principle of least action and show that Newton's laws are a consequence of it. After discussing the conservation laws, we will then proceed to prove Kepler's laws using a few different methods: geometrical, using Newton's methods, using the Laplace-Runge-Lenz vector and finally using the least action principle.

The maths in this chapter is much more engaging than the previous two chapters. However, we believe that at the end, a unified vision of celestial motion will appear that is intellectually extremely satisfying.

3.1 THE PRINCIPLE OF LEAST ACTION

If we find ourselves at the very top of a mountain we experience a very peculiar position: we are in the only flat place. In three dimensions, we can model the mountain by a function $z(x, y)$ of two variables $z = f(x, y)$. The top of the mountain is special because if we move just a tiny bit in the x or y direction, the coordinate z does not change much. In effect, a technique to find the maximum of a function consists of finding where the *first derivative* is zero, i.e. what point shows the property that if we move a tiny bit we do not lose altitude. Any other point that is not a maximum (or a minimum like a valley) does not have this property.

Now we state a law which, as we previously specified many times, has to be accepted as true until an *observation* contradicts it. The **principle of least action**: Given two points in space P and Q, a particle will move from P to Q along a trajectory which *minimizes* the following integral:

$$S = \int_{t=t_1}^{t=t_2} \mathcal{L} \, dt \qquad (3.1)$$

where $\mathcal{L}$ is a special function called *Lagrangian* from the Italian mathematician Joseph-Louis Lagrange (actually born in Turin (Italy) with the name Giuseppe Lodovico Lagrangia). Notice that all the possible paths between P and Q have to be traveled in the same time ($t_2 - t_1$) (see fig. 3.2).

Let us now study what conditions the function $\mathcal{L}$ must satisfy in order to have the path $\overline{PQ}$ be the path of least action. Let us now restrict ourselves to the simple unidimensional case in which the function $\mathcal{L}$ is a function of

---

[1]More accurately it should be stated as the *principle of stationary action*.

FIGURE 3.1 Joseph-Louis Lagrange, born Giuseppe Luigi Lagrangia, was an Italian mathematician and astronomer.

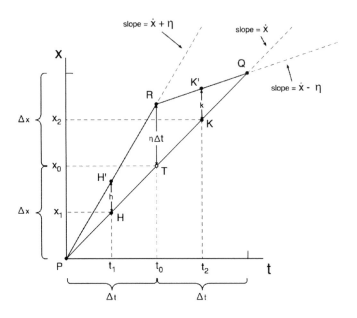

FIGURE 3.2 Geometry of the principle of least action .

the space coordinate x and its derivative dx/dt. Let us indicate the derivative dx/dt with the symbol[2] $\dot{x}$. We make the following approximations:

1. The function $\mathcal{L} = \mathcal{L}(x, \dot{x})$ can be approximated with a Taylor's expansion limited to the first order.

2. All the segments in fig. 3.2 are infinitesimal.

We already encountered a power series expansion in eq. 1.41 where a function $P(x)$ is approximated at a certain point $x = x_0$ with an infinite series of polynomials of increasing order whose coefficients are the derivatives calculated at $x = x_0$. A Taylor's expansion is exactly this. Our function $\mathcal{L} = \mathcal{L}(x, \dot{x})$ is a function of two variables x and $\dot{x}$. In this case, the Taylor expansion limited to the first order around the point R where $x = x_0$ and $t = t_0$, is the value of the function at the point T ($= \mathcal{L}(0)$) times the infinitesimal shift in $x (= \eta\Delta t)$ times the partial derivative of $\mathcal{L}$ with respect to x evaluated at $t = t_0$ plus the infinitesimal change in slope $\dot{x}(= \eta)$ times the partial derivative of $\mathcal{L}$ with respect to $\dot{x}$ also evaluated at $t = t_0$:

$$\mathcal{L} = \mathcal{L}(0) + \Delta x \frac{\partial \mathcal{L}}{\partial x}\Big|_{t=t_0} + \Delta \dot{x} \frac{\partial \mathcal{L}}{\partial \dot{x}}\Big|_{t=t_0} \qquad (3.2)$$

where the symbol $\partial \mathcal{L}/\partial x\big|_{t=t_0}$ means the derivative of $\mathcal{L}$ with respect to x,

---

[2]This is Newton's convention and it is more compact than the dx/dt symbol by Leibniz.

evaluated at $t = t_0$, while keeping $\dot{x}$ constant and $\partial\mathcal{L}/\partial\dot{x}|_{t=t_0}$ means the derivative of $\mathcal{L}$ with respect to $\dot{x}$, evaluated at $t = t_0$, while keeping x constant.

The integral 3.1 can be written by taking the average of the function $\mathcal{L}$.

$$S = \int_{t=t_1}^{t=t_2} \mathcal{L}\ dt = \mathcal{L}_{\text{average}}(t_2 - t_1) \tag{3.3}$$

Since we are dealing with a straight line, the average value is exactly in the middle of the various segments.

We now study the path of minimal action $\overline{PQ}$ and we consider a simple infinitesimal variation of this path by tracing two straight lines $\overline{PR}$ and $\overline{RQ}$ (fig. 3.2). Our particle can go from P to Q either along the straight line $\overline{PQ}$, which is also the minimal action line, or along a path from $\overline{PR}$ and then $\overline{RQ}$. It is important to underline that the time taken along the path $\overline{PQ}$ is the same as the total time to travel the path $\overline{PQ} + \overline{RQ}$. This means that the particle will travel first at a slightly higher speed along $\overline{PR}$ than at a slightly lower speed along $\overline{RQ}$ than along the path $\overline{PQ}$.

Let's estimate the action integral 3.1 by using eq. 3.2. The average value of the action S_{PQ} along $\overline{PQ}$ is given by the value of $\mathcal{L}$ at the point T, i.e. $\mathcal{L}(x_0,\dot{x}_0)$, times the total time interval to go from P to Q, $2\Delta t$:

$$S_{PQ} \approx 2\Delta t \mathcal{L}(x_0,\dot{x}_0) \tag{3.4}$$

Before calculating the two action integrals along $\overline{PR}$ and $\overline{RQ}$, let's estimate the two variations $h = \overline{HH'}$ and $k = \overline{KK'}$. In order to do that, we need to first calculate the value of the segment $\overline{TR}$. This is easily accomplished by writing explicitly the equation of the two straight lines $\overline{PR} = (\dot{x} + \eta)t$ and $\overline{PQ} = \dot{x}t$ where η is a small increase in the slope of the straight line $\overline{PQ}$. The segment $\overline{TR}$ is given by:

$$\overline{TR} = (\dot{x} + \eta)t_0 - \dot{x}t_0 = \eta t_0 = \eta \Delta t \tag{3.5}$$

In fig. 3.2 we have that, by construction, the point T divides exactly $\overline{PQ}$ in half, i.e. $\overline{PT} = \overline{TQ} = \frac{1}{2}\overline{PQ}$. Similarly, $\overline{PH} = \frac{1}{2}\overline{PT}$ and $\overline{TK} = \frac{1}{2}\overline{TQ}$. This means that the triangles $\triangle PRT$ and $\triangle PH'H$ are similar. If they are similar, then we must have that $h = \frac{1}{2}\overline{TR} = \frac{1}{2}\eta\Delta t$. The term $\dot{x} - \dot{x}_1$ must be equal at the variation of slope which is $k = \eta$.

We now have all the ingredients to estimate the action integral at point H', which can be obtained by Taylor expansion around the point H, remembering that now we are averaging over the line $\overline{PR}$ for a time interval Δt:

$$S_{PR} \approx \Delta t \left[\mathcal{L}(x_1,\dot{x}_1) + h\frac{\partial\mathcal{L}}{\partial x}\Big|_{t=t_1} + k\frac{\partial\mathcal{L}}{\partial\dot{x}}\Big|_{t=t_1} \right]$$
$$= \Delta t \left[\mathcal{L}(x_1,\dot{x}_1) + \frac{1}{2}\eta\Delta t\frac{\partial\mathcal{L}}{\partial x}\Big|_{t=t_1} + \eta\frac{\partial\mathcal{L}}{\partial\dot{x}}\Big|_{t=t_1} \right] \tag{3.6}$$

In a similar way, by averaging along the segment $\overline{RQ}$ we have that the action integral at point K' is:

$$
\begin{aligned}
S_{RQ} &\approx \Delta t \left[\mathcal{L}(x_2, \dot{x}_2) + h \frac{\partial \mathcal{L}}{\partial x}\Big|_{t=t_2} - k \frac{\partial \mathcal{L}}{\partial \dot{x}}\Big|_{t=t_2} \right] \\
&= \Delta t \left[\mathcal{L}(x_2, \dot{x}_1) + \frac{1}{2}\eta \Delta t \frac{\partial \mathcal{L}}{\partial x}\Big|_{t=t_2} - \eta \frac{\partial \mathcal{L}}{\partial \dot{x}}\Big|_{t=t_2} \right]
\end{aligned}
\tag{3.7}
$$

where we used the fact that $\dot{x}_2 = \dot{x}_1$, i.e. the slopes at points H and K are equal to the value $\Delta x / \Delta t$.

We are now ready to compare eqs. 3.4, 3.6 and 3.7 which express an estimate of the action integral along the corresponding segments. We now add 3.6 and 3.7 to obtain the action integral along the total trajectory $\overline{PR}$ plus $\overline{RQ}$:

$$
\begin{aligned}
S_{PR} + S_{RQ} &\approx \Delta t \left[\mathcal{L}(x_1, \dot{x}) + \mathcal{L}(x_2, \dot{x}) \right] \\
&+ \frac{1}{2}\eta \Delta t^2 \left[\frac{\partial \mathcal{L}}{\partial x}\Big|_{t=t_1} + \frac{\partial \mathcal{L}}{\partial x}\Big|_{t=t_2} \right] \\
&+ \eta \Delta t \left[\frac{\partial \mathcal{L}}{\partial \dot{x}}\Big|_{t=t_1} - \frac{\partial \mathcal{L}}{\partial \dot{x}}\Big|_{t=t_2} \right]
\end{aligned}
\tag{3.8}
$$

The first terms on the right side of eq. 3.8 are averages of the actions at their midpoints corresponding to $t = t_0$. The last term is the difference of the function between two points that are close in time which is approximated by the derivative with respect to time of the function, times the infinitesimal time interval Δt. We can therefore write:

$$
\begin{aligned}
S_{PR} + S_{RQ} &\approx \Delta t \left[2\mathcal{L}(x_0, \dot{x}) + \eta \Delta t \frac{\partial \mathcal{L}}{\partial x}\Big|_{t=t_0} - \eta \Delta t \frac{d}{dt}\frac{\partial \mathcal{L}}{\partial \dot{x}}\Big|_{t=t_0} \right] \\
&= \Delta t \left[2\mathcal{L}(x_0, \dot{x}) + \eta \Delta t \left(\frac{\partial \mathcal{L}}{\partial x}\Big|_{t=t_0} - \frac{d}{dt}\frac{\partial \mathcal{L}}{\partial \dot{x}}\Big|_{t=t_0} \right) \right]
\end{aligned}
\tag{3.9}
$$

Comparing eq. 3.4 with eq. 3.9 we see that the two actions, in order to be equal, must satisfy the following equation:

$$
\frac{\partial \mathcal{L}}{\partial x}\Big|_{t=t_0} - \frac{d}{dt}\frac{\partial \mathcal{L}}{\partial \dot{x}}\Big|_{t=t_0} = 0
\tag{3.10}
$$

Eq. 3.10 is known as the **Euler-Lagrange** equation. More in general, the Euler-Lagrange is a set of equations referred to *generic* coordinates q and their time derivatives $\dot{q}$. Therefore, a more general Euler-Lagrange equation can be written as:

$$
\frac{\partial \mathcal{L}}{\partial q_i} - \frac{d}{dt}\frac{\partial \mathcal{L}}{\partial \dot{q}_i} = 0
\tag{3.11}
$$

where we can arbitrarily choose the coordinates q_i, usually referred to as *generalized coordinates*. The generalized coordinates refer to the coordinates describing the configuration of the system. These coordinates must uniquely define the configuration of the system relative to a reference configuration. The generalized velocities $\dot{q}_i$ are the time derivatives of the generalized coordinates of the system. An example of a generalized coordinate is the angle that locates a planet moving on its orbit around the Sun.

We can choose to describe the motion of a planet either by its Cartesian coordinates x, y and associated velocities $\dot{x}, \dot{y}$ or through the use of polar coordinates r, θ and associated *velocities* $\dot{r}, \dot{\theta}$. It is a matter of convenience which coordinate we want to use.

We have suffered quite a bit during our last proof, but we now show how powerful and more fundamental the last equation is with respect to Newton's laws. Eq. 3.10 tells us that we have a function $\mathcal{L}(x, \dot{x})$ of a variable x and its derivative with respect to time $\dot{x}$ which, if it is a solution, the particle will travel along the "real" trajectory. We therefore need to specify this function $\mathcal{L}(x, \dot{x})$ called **Lagrangian**. It turns out that such a function is defined as:

$$\mathcal{L}(x, \dot{x}) = T - V \tag{3.12}$$

where T and V are, respectively, the kinetic and potential energy of the particle. Eq. 3.12 establishes an empirical fact, i.e. a law. Nobody knows exactly why the Lagrangian must be as in eq. 3.12, but if we build the Lagrangian as the difference between kinetic and potential energy, then the minimization of the action integral with the function of eq. 3.12 describes the motion of the particle.

3.1.1 Conservation Laws

There are three fundamental conservation laws that in Newtonian mechanics, need to be given as laws. We will now show that these conservation laws are actually derived from symmetries in nature. The first thing to do is to define the energy terms in eq. 3.12. The kinetic energy of a particle of mass m traveling with a *speed* v is:

$$T = \frac{1}{2}m\dot{x}^2 = \frac{1}{2}mv^2 \tag{3.13}$$

Let us now assume a simple form for the function V. We say that the potential energy $V(x)$ is a *scalar* function of the coordinate x. In this case, the Lagrangian of a particle of mass m subject to a potential energy given by a scalar function $V = V(x)$ is:

$$\mathcal{L}(x, \dot{x}) = \frac{1}{2}m\dot{x}^2 - V(x) \tag{3.14}$$

Let us calculate the output of the Euler-Lagrange equation with a Lagrangian given by eq. 3.14. We have that:

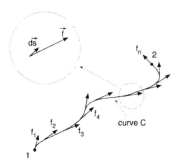

FIGURE 3.3 Definition of work done by the force f along the curve C.

$$\frac{\partial \mathcal{L}}{\partial x} = \frac{dV}{dx}$$
$$\frac{\partial \mathcal{L}}{\partial \dot{x}} = m\dot{x} \tag{3.15}$$
$$\frac{d}{dt}\frac{\partial \mathcal{L}}{\partial \dot{x}} = m\ddot{x} = ma$$

Eq. 3.15 is Newton's equation $F = ma$ providing that the force F can be derived as the derivative of a scalar potential $V(x)$.

$$F = m\ddot{x} = -\frac{dV}{dx} \tag{3.16}$$

What kind of forces are generated by the derivative of a scalar function?

First, we need to introduce the concept of *work* . A force $\overrightarrow{F}$ does **work** if there is a displacement of the point of application of the force along the force direction. In fig. 3.15 a force is applied along the curve C from the point 1 to the point 2. According to the definition of work, an infinitesimal amount of work dW is obtained from the scalar product between the infinitesimal displacement $d\overrightarrow{s}$ and the force $\overrightarrow{f}$:

$$dW = \overrightarrow{f} \cdot d\overrightarrow{s} \tag{3.17}$$

Eq. 3.17 tells us that the work is a scalar quantity (number). If we want to find the total work done between the points 1 and 2 we need to sum over all the infinitesimal displacement $d\overrightarrow{s}$. The total work is therefore:

$$W_{12} = \int_{1}^{2} \overrightarrow{f} \cdot d\overrightarrow{s} \tag{3.18}$$

Now, let's look back at eq. 3.16 and assume that the motion is happening only in one dimension x. The scalar product becomes simply the product fdx and we have:

$$W_{12} = \int_1^2 fdx = \int_1^2 -\frac{dV}{dx}dx = V(1) - V(2) \tag{3.19}$$

where we have used the Fundamental Theorem of Calculus 1.31. Eq. 3.19 tells us that the class of forces obtained as the derivative of a scalar function are very special: the work done by these forces does not depend on the particular path but only on the difference between the final and initial points. These special forces are called **conservative forces**.

We can be a bit more specific about the function $V = V(x)$, called **potential energy**. Without defining energy precisely, for the moment, let's assume we know what energy is, i.e. the capacity of doing work. Then the Lagrangian 3.12 is the **difference** between the kinetic energy and the potential energy where the potential energy is that form of energy that depends only on the position of a body with respect to another. Think about the potential of doing damage that a brick located on the 20th floor of a building has. If dropped from the height of 1 cm it can barely scratch your hand, but from the height of 60 meters it will easily kill you. Two very important examples of conservative forces are electric forces between two electric charges and gravitational attraction between two masses.

So why does nature decide that the path that minimizes the difference between kinetic and potential energy is the path that the system will follow? Nobody knows why! The reality is that it works very well and has many useful applications. One of the most elegant applications regards the symmetries of the physical laws.

We have seen that the Lagrangian 3.12 is a function of the coordinate x and its time derivative $\dot{x}$. In general, the Lagrangian can be also explicitly a function of time t, $\mathcal{L} = \mathcal{L}(x, \dot{x}, t)$. Now we discuss three different ways to perform a specific experiment. Suppose we describe a system with a certain Lagrangian $\mathcal{L}$ and we require that the same system is described with the same Lagrangian but in a different location on the Earth. This means that if a certain physical law is valid, for example, in Rome, then it must be valid also in New York. We are asking under what conditions such a request can be satisfied. In other words we request that if we change x into $x + \epsilon$, the Lagrangian does not change. There is a very special situation that satisfies this requirement: the Lagrangian does not contain the term x explicitly. In this case we are free to choose whatever x we want because it is not explicitly contained in $\mathcal{L}$. Now, if the coordinate x is not present in $\mathcal{L}$, then the corresponding derivative is equal to zero and the Euler-Lagrange equation becomes:

$$\frac{\partial \mathcal{L}}{\partial x} - \frac{d}{dt}\frac{\partial \mathcal{L}}{\partial \dot{x}} = 0$$
$$\frac{\partial \mathcal{L}}{\partial x} = 0$$
$$\frac{d}{dt}\frac{\partial \mathcal{L}}{\partial \dot{x}} = 0 \tag{3.20}$$
$$\frac{\partial \mathcal{L}}{\partial \dot{x}} = constant$$

In the case of a free particle in absence of any external force, we have that the Lagrangian is simply $\mathcal{L} = \frac{1}{2}m\dot{x}^2$ and the last equation in 3.20 becomes:

$$\frac{\partial}{\partial \dot{x}}\left(\frac{1}{2}m\dot{x}^2\right) = m\dot{x} = constant \tag{3.21}$$

The quantity $m\dot{x}$ is the momentum and eq. 3.21 tells us that the momentum is conserved. Alternatively, we say that since Lagrangian is invariant for translations in space (i.e. we can choose whatever x we want since the Lagrangian does not contain x explicitly), then momentum is conserved.

When the Lagrangian is independent from changes of one of its variables, we say that the Lagrangian is *symmetric* with respect to that specific coordinate. This concept is in complete analogy with the more familiar use of the word "symmetric". For example, we say that a square is symmetric under rotation of $90°$ because under such rotation the square returns identical to itself. On the other hand, we cannot say that a square is symmetric under *any* rotation, because a small rotation will change its orientation. So a square is not symmetrical under *continuous* rotation, meaning that it does change for small rotations. A circumference is an example of *continuous* symmetry under rotation. No matter how small the angle of rotation, the circumference does not change.

The symmetry of the Lagrangian under continuous transformations generates conserved quantities. We have seen that under translations the momentum is conserved. What about if our Lagrangian is symmetric under continuous rotations in space? It means that our system does not change if we rotate everything in the Universe. For this particular symmetry, we find that expressing the Lagrangian in Cartesian coordinates is not convenient. First, since we are rotating, we need a plane and so we need to express the generic Lagrangian as $\mathcal{L}(x, \dot{x}, y, \dot{y})$ because now we work on the plane x, y. We have to transform the coordinates from x, y to r, θ. Using eq. 1.27 for finding the derivative of a product of two functions and eq. 2.8 for the rule to rotate a coordinate system, we have:

$$x = r \cos \theta$$
$$y = r \sin \theta$$
$$\dot{x} = \dot{r} \cos \theta - r\dot{\theta} \sin \theta$$
$$\dot{y} = \dot{r} \sin \theta + r\dot{\theta} \cos \theta$$

To write down the Lagrangian, we need to express the kinetic energy T and the potential energy V in the new coordinate system. The kinetic energy is:

$$T = \frac{1}{2}m(\dot{x}^2 + \dot{y}^2) = \frac{1}{2}m\left[\left(\dot{r}\cos\theta - r\dot{\theta}\sin\theta\right)^2 + \left(\dot{r}\sin\theta + r\dot{\theta}\cos\theta\right)^2\right]$$
$$= \frac{1}{2}m\left(\dot{r}^2 + r^2\dot{\theta}^2\right)$$

(3.22)

Now let us suppose our system is subject to a conservative force such that the potential energy is only a function of r, $V = V(r)$. The total Lagrangian in polar coordinates will be:

$$\mathcal{L}(r, \dot{r}, \theta, \dot{\theta}) = \frac{1}{2}m\left(\dot{r}^2 + r^2\dot{\theta}^2\right) - V(r) \qquad (3.23)$$

Looking at eq. 3.23 we immediately notice that there is no explicit dependency on the coordinate θ. So this Lagrangian is already symmetric with respect to rotations. In complete analogy with the case of invariance with translations, where the Lagrangian did not have explicit dependence on x, we now have no explicit dependence on θ. This means that the following equation must hold:

$$\frac{\partial}{\partial\theta}\frac{1}{2}m\left[\left(\dot{r}^2 + r^2\dot{\theta}^2\right) - V(r)\right] = mr^2\dot{\theta} = constant \qquad (3.24)$$

Eq. 3.24 shows that the quantity $mr^2\dot{\theta}$ is constant, i.e. is conserved. This quantity is called the **angular momentum**.

The last symmetry we want to discuss concerns the invariance of the Lagrangian with respect to time shifts. In other terms, if an experiment is performed today and we obtain a certain result, if we repeat the same experiment under exactly the same conditions, we must find exactly the same result. Let's assume now that the Lagrangian depends only on one coordinate, its derivative, and the time t, $\mathcal{L}(x, \dot{x}, t)$. The condition of time invariance corresponds imposing that the derivative of the Lagrangian with respect to time is zero. Using the chain rule (eq. 1.30), we have:

$$\frac{d}{dt}\mathcal{L}(x,\dot{x},t) = \frac{\partial \mathcal{L}}{\partial x}\frac{dx}{dt} + \frac{\partial \mathcal{L}}{\partial \dot{x}}\frac{d\dot{x}}{dt} + \frac{\partial \mathcal{L}}{\partial t}$$
$$= \frac{d}{dt}\left(\dot{x}\frac{\partial \mathcal{L}}{\partial \dot{x}} - \mathcal{L}\right) = 0 \tag{3.25}$$

Eq. 3.25 tells us that in order to have the Lagrangian not explicitly depend on time (i.e. its derivative with time equal to zero) the term in parenthesis must be equal to a constant:

$$\dot{x}\frac{\partial \mathcal{L}}{\partial \dot{x}} - \mathcal{L} = constant \tag{3.26}$$

Using 3.21, we have that:

$$m\dot{x}^2 - \mathcal{L} = 2T - (T - V) = T + V = constant \tag{3.27}$$

Time invariance of the Lagrangian requires that the quantity $T + V$ is constant. This term is the mechanical energy and therefore eq. 3.27 requires the **conservation of mechanical energy** defined as the sum of the kinetic energy plus the potential energy. The three conservation laws that we just obtained are a consequence of one of the most beautiful theorems in mathematics (and physics), called **Noether's theorem**, named after the German mathematician Amalie Emmy Noether, probably the most important woman in the history of mathematical physics.

Let's stop for a moment and reflect on what we have just stated. In addition to Newton's method of solving the motion of bodies subject to external forces and therefore using vectors, we have found that an alternative, more powerful, description can be obtained through the principle of least action[3]. The novelty is that in order to use the machinery of the Euler-Lagrange equations, we need to be able to write down the kinetic and potential energies which are scalar and not vector quantities. One major advantage of Lagrangian formulation is that it is extensible to other areas of physics like quantum mechanics and optics. Finally, the symmetry properties of the Lagrangian give immediately conserved quantities as we have seen with momentum, angular momentum, and energy.

3.1.2 Newtonian and Lagrangian Problem Solving

We now show how to solve a simple problem with Newton's approach, and then we show how the Lagrangian produces identical results. Let us assume we have a mass m attached to a spring of negligible mass. Suppose we have verified that the spring, when compressed or extended, always reacts with a force opposed to the compression (extension) and proportional to the amount

---

[3]In reality, we do not strictly need that the action is a minimum and therefore we should request that the action is *stationary*.

FIGURE 3.4 Amalie Emmy Noether.

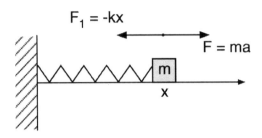

$F_1 = -kx$

$F = ma$

m

x

FIGURE 3.5 Mass m at position x on a frictionless table constrained by a spring of elastic constant k.

of shift, i.e. the force from the spring is $F = -kx$ where k is a constant typical of the spring and the minus sign means that the force is opposed to the compression (extension).

In fig. 3.5 a mass m is on a frictionless table attached to a wall by a massless spring of elastic constant k, i.e. capable of producing a force always opposed to any pull or push. We assume that the strength of this reaction force from the spring is linearly proportional to the amount of stretch or compression, and opposed, i.e. the spring pushes away if compressed, or pulls back if extended. If we call the spring reaction force F_1, we can write that the reaction force is $F_1 = -kx$. Newton's law then tells us that the mass will be subject to the following equation:

$$F = ma = -kx \tag{3.28}$$

Note that eq. 3.28 tells us that the mass m has zero force when in $x = 0$. If we express the acceleration a as the second derivative of the space coordinate x, we have:

$$\frac{d^2x}{dt^2} = -\frac{k}{m}x \tag{3.29}$$

Eq. 3.29 is an example of a differential equation of the second order, meaning that we are looking for a function $x = x(t)$ that differentiated twice gives the same function $x(t)$ multiplied by a factor $-\frac{k}{m}$. An example of a function that once differentiated twice comes back to itself is the function sine whose derivatives are is listed in table 1.1. Let's try a solution $x(t) = A\sin(\omega t + \phi)$. Let us use the notation $\dot{x} \equiv \frac{dx}{dt}$:

$$
\begin{aligned}
x &= A\sin(\omega t + \phi) \\
\dot{x} &= \omega A\cos(\omega t + \phi) \\
\ddot{x} &= -\omega^2 A\sin(\omega t + \phi)
\end{aligned}
\tag{3.30}
$$

Let's now insert 3.30 into 3.29:

$$\ddot{x} = -\omega^2 A \sin(\omega t + \phi) = -\frac{k}{m} A \sin(\omega t + \phi)$$
$$-\omega^2 = -\frac{k}{m} \tag{3.31}$$

so, the solution to eq. 3.29 is:

$$x(t) = A \sin\left(\sqrt{\frac{k}{m}} t + \phi\right) \tag{3.32}$$

which describes a sinusoidal oscillation of the mass m at a characteristic frequency $\omega = 2\pi\nu = \sqrt{\frac{k}{m}}$.

Let us solve the same problem using the Lagrangian formalism. First, we need to identify the kinetic and the potential energies. The kinetic energy of the mass m is $T = \frac{1}{2}m\dot{x}^2$. In order to find the potential energy we use a generalization of eq. 3.19 that tells us that the potential energy is the integral of the force along the trajectory. Since we are considering a simple case where the mass is constrained along the x axis, the potential $V(x)$ is obtained:

$$V(x) = \int \frac{dV}{dx} dx = -\int F dx = \int -kx \, dx = \frac{1}{2}kx^2 \tag{3.33}$$

so the Lagrangian is:

$$\mathcal{L} = T - V = \frac{1}{2}m\dot{x}^2 + \frac{1}{2}kx^2 \tag{3.34}$$

We now use the Euler-Lagrange equations 3.10:

$$\frac{\partial \mathcal{L}}{\partial x} = kx$$
$$\frac{\partial \mathcal{L}}{\partial \dot{x}} = m\dot{x}$$
$$\frac{d}{dt}\frac{\partial \mathcal{L}}{\partial \dot{x}} = m\ddot{x} \tag{3.35}$$
$$m\ddot{x} = -kx$$

and so we obtain the equation 3.29.

You might ask: "why use the Lagrangian formalism to obtain the same result as the Newtonian formalism?"; or equivalently, "is it worth the complication of the Lagrangian formalism?" There are situations where using the Newtonian formalism is not as simple as in the case of the harmonic oscillator. For example, a system of 3 springs and two masses as in fig. 3.6 is quite complicated to treat with Newton formalism. With the Lagrangian we need to write the correct kinetic and potential energies. The kinetic energy of the system is

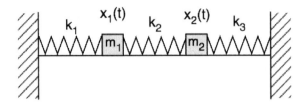

FIGURE 3.6 Two masses m_1 and m_2 on a friction less table constrained by three springs of elastic constant k_1, k_2 and k_3.

just the sum of the two kinetic energies of each mass $T = \frac{1}{2}m_1\dot{x}_1^2 + \frac{1}{2}m_2\dot{x}_2^2$; the potential energy is the sum of three terms because we have three springs contributing to it: $V(x) = \frac{1}{2}k_1x_1^2 + \frac{1}{2}k_2(x_2 - x_1)^2 + \frac{1}{2}k_3x_3^2$. It is now relatively easy, and we leave it to the reader to check that the equation of motions are:

$$\begin{aligned} m_1\ddot{x}_1 &= -k_1x_1 + k_2(x_2 - x_1) \\ m_2\ddot{x}_2 &= -k_2(x_2 - x_1) - k_3x_2 \end{aligned} \tag{3.36}$$

We will not discuss or solve eq. 3.36 because it is outside the objectives of this book. The example was briefly discussed to show the power of the Lagrangian formalism over the Newtonian for the specific case of a coupled harmonic oscillator.

3.2   KEPLER'S LAWS

We now turn our attention to Johannes Kepler, born in Weil der Stadt (Germany) in 1571. Kepler was a mathematician and astronomer who carefully studied the motion of the Moon and the planets known at the time (Earth, Mercury, Venus, Mars, Jupiter and Saturn). We have already mentioned that he worked as an assistant to the Danish astronomer Tycho Brahe, who collected a huge amount of naked eye observations of planet positions. Kepler happened to be the right person at the right time to analyze this data. After years of attempts, Kepler came up with three laws for the motion of the planets. Let us remind the reader that these laws are *empirical* in the sense that they best interpret the data known at that time. Until there is experimental data not explained, scientists assume that the laws are universally valid. Today we know that Kepler's laws (and Newton's laws) are not valid, but need to be corrected using Einstein's general relativity. In this book we consider the celestial mechanics only from a classical point of view, leaving the more advanced treatment of general relativity for specialized textbooks.

Let's go back to Kepler and his laws. They can be written as follows:

FIGURE 3.7 Johannes Kepler was a German astronomer and mathematician most famous for his 3 laws governing the motion of planets around the Sun.

Kepler 1st law: Planets revolve around the Sun on elliptical orbits where the Sun lies at one of the foci.

Kepler 2nd law: The segment (radius vector) connecting the planets to the Sun sweeps out equal areas in equal time intervals.

Kepler 3rd law: The square of the orbital period of each planet is proportional to the cube of its orbital major axis.

Let's now imagine that we go back to Kepler's time with our modern understanding of mathematics and see how we can re-obtain his beautiful laws and the connection with Newton's laws[4]. We will now review quite a lot of math and the journey will be quite engaging, but it will be worth the effort because at the end we will have a deeper understanding of the relationships between geometry, calculus, and physics.

3.2.1 Theory of Conic Sections

We start our long journey with the mathematics of conic sections, i.e. the special class of functions obtained by intersecting a cone with a plane. A deep understanding of conic sections will allow us to study the orbital motion not only for planets but also for artificial satellites. We will approach the conic sections with some advanced math, namely matrix representation of conic sections, which gives us the opportunity to introduce higher-level math. This is not just *per se*, but it is used to show some inner symmetries and beautiful connections with other fields in math and physics. Our brief discussion of eigenvalues and eigenvectors, for example, can serve as a springboard to the math used in quantum mechanics.

In fig. 3.8 we see that the circle, ellipse, parabola, and hyperbola are all obtained by "sectioning" a double cone with a plane, thus the name *conic sections*. There are also special cases when the plane intersection results in a point, a line, or in a couple of crossing lines. In this case we say that the conic section is *degenerate*.

The general equation of a conic is given by:

$$Ax^2 + 2Bxy + Cy^2 + 2Dx + 2Ey + F = 0 \qquad (3.37)$$

where the factors 2 are inserted to simplify later calculations. There is a special quantity, called the **discriminant**, which allows us to classify the conic sections depending on its sign. The discriminant of the quadratic form 3.37 is given by:

$$\Delta_2 = B^2 - AC \qquad (3.38)$$

We have that:

1. If $\Delta < 0$ and $B = 0$ and $A = C$, then the conic is a **circle**.

---

[4] As we mentioned, we are not considering general relativity.

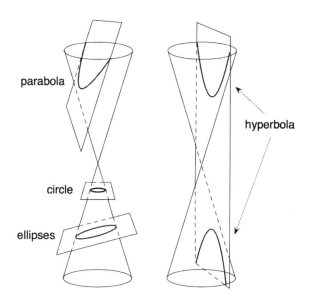

FIGURE 3.8   Conic sections obtained by intersecting a plane with a double cone.

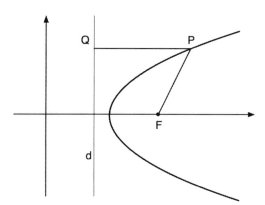

FIGURE 3.9   A generic conic section defined as the locus of the point P such that $\overline{PF}/\overline{PQ} = e$, where e is the eccentricity.

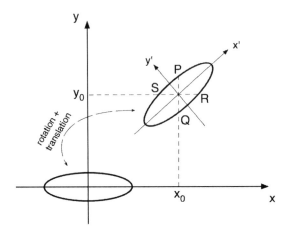

FIGURE 3.10 Generic ellipse described by eq. 3.37 with center at x_0, y_0 and axes x' and y' rotated with respect to the x, y coordinate system.

2. If $\Delta < 0$, then the conic is an **ellipse**.

3. If $\Delta = 0$, then the conic is a **parabola**.

4. If $\Delta > 0$, then the conic is a **hyperbola**.

A conic section is defined as the locus of a point P that moves in the plane of a fixed point F, called the focus, and a fixed line d, called the conic section directrix (with F not on d) such that the ratio of the distance of P from F to its distance from d is a constant e, called the eccentricity (see fig. 3.9). The form 3.37 describes a conic section where the center is on a generic point on the Cartesian plane and the axis can have any orientation (see fig. 3.10).

It is possible to find a coordinate transformation such that the conic is expressed in a new Cartesian coordinate system x', y', such that the center is at the origin and the axes are parallel to x' and y' axes. If we succeed then the conic section is written in its **canonical form**.

Let's state what characteristics such a coordinate transformation must have. First, vectors must not change their length, and second, the angles between vectors must not change. We have that the scalar product of two vectors $\vec{u} \cdot \vec{v}$ defines both the lengths and the angle between two vectors $\vec{u}$ and $\vec{v}$. So we say that a coordinate transformation that *preserves* the scalar product will make sure that we can shift and rotate our conic sections without changing their shape (also termed as *rigid transformations* imagining that we are moving a rigid object). Transformations that preserve the scalar product are called **orthogonal transformations**.

In section 2.1.1 we have shown that a coordinate transformation, specifically a rotation of an angle α, can be represented with a square table of 2×2

numbers in a 2D coordinate system called a **matrix**. We now show how the matrix formalism can clearly treat the problem of rigid transformations that will allow us to express a conic section well centered in a coordinate system. From now on, instead of talking about *transformations*, we will talk about *matrices* being the two concepts absolutely equivalent. A matrix representing an orthogonal transformation is called **orthogonal matrix**. The *transpose* of a matrix A is a matrix where the rows are exchanged with the columns and is indicated with the symbol A^T:

$$A = \begin{pmatrix} a & b \\ c & d \end{pmatrix}, \quad A^T = \begin{pmatrix} a & c \\ b & d \end{pmatrix} \tag{3.39}$$

Let us give a few definitions:

Definition: Let A be a square matrix of size n. A is a **symmetric**[5] matrix if $A^T = A$.

Definition: A matrix P is said to be **orthogonal** if its columns are mutually orthogonal.[6]

Definition: A matrix P is said to be **orthonormal** if its columns are unit vectors[7] and P is orthogonal.

Definition: A matrix D is said to be **diagonal** if all its elements are zero except the element on the diagonal. Matrix A in 3.39 would be diagonal if $b = c = 0$.

Definition: The determinant of a matrix A, 2×2 size, is a number calculated in the following way:

$$|A| \equiv det(A) = det \begin{pmatrix} a & b \\ c & d \end{pmatrix} = (ad - bc) \tag{3.40}$$

The determinant of a 3×3 matrix is a bit more complex. It can be broken down to the sum of 3 determinants of 2×2 matrices. The general formula is given by:

$$det \begin{pmatrix} a & b & c \\ d & e & f \\ g & h & i \end{pmatrix} = a \, det \begin{pmatrix} e & f \\ h & i \end{pmatrix} - b \, det \begin{pmatrix} d & f \\ g & i \end{pmatrix} + c \, det \begin{pmatrix} d & e \\ g & h \end{pmatrix} \tag{3.41}$$

or equivalently:

$$det \begin{pmatrix} a & b & c \\ d & e & f \\ g & h & i \end{pmatrix} = a(ei - fh) - b(di - fg) + c(dh - eg) \tag{3.42}$$

---

[5]Clearly matrix A in eq. 3.39 is not symmetric. To be symmetric it must have $b = c$.

[6]Remember that vectors are orthogonal if their scalar product is zero. In 2D we would say that they are *perpendicular*.

[7]A unit vector is a vector whose length is equal to 1.

It is possible to express the general conic section form of eq. 3.37 in matrix form as follows:

$$(x \quad y) \begin{pmatrix} A & B \\ B & C \end{pmatrix} \begin{pmatrix} x \\ y \end{pmatrix} + 2(D \quad E) \begin{pmatrix} x \\ y \end{pmatrix} + F = 0 \tag{3.43}$$

where the matrix Q is:

$$Q = \begin{pmatrix} A & B \\ B & C \end{pmatrix} \tag{3.44}$$

We read eq. 3.43 from left to right as follows: the $(x \quad y)$ is a *row* vector with coordinates x, y. When multiplied with a column vector like, for example, $(x \quad y) \begin{pmatrix} x \\ y \end{pmatrix}$ gives the number $x^2 + y^2$. It is another way to indicate the scalar product. The matrix Q multiplied by the vector $\begin{pmatrix} x \\ y \end{pmatrix}$ is then multiplied by the vector $(x \quad y)$ to give $Ax^2 + 2Bxy + Cy^2$. The remaining terms are obtained by multiplying the two vectors $2(D \quad E)$ with $\begin{pmatrix} x \\ y \end{pmatrix}$ to give $2Dx + 2Ey$, with F being just a number.

An equivalent matrix form, using a 3×3 matrix A_Q is:

$$(x \quad y \quad 1) \begin{pmatrix} A & B & D \\ B & C & E \\ D & E & F \end{pmatrix} \begin{pmatrix} x \\ y \\ 1 \end{pmatrix} = 0 \tag{3.45}$$

If we call $\vec{z} = (x \quad y \quad 1)$, then eq. 3.37 can be written in a very compact form:

$$z^T A_Q \, z = 0 \tag{3.46}$$

With ref. to fig. 3.10, we see that we can center and align the axis of the conic section if we go from the system x, y to the system x', y'. This operation can be obtained by first shifting the coordinate system x, y to have a new origin in x_0, y_0 corresponding to the center of the conic (in this case an ellipse) and a rotation of the proper angle to align the coordinate axis to the ellipse axis.

Before shifting the coordinate system, let us find the center of the conic (in this case an ellipse) taking into account that a parabola does not have a center. Looking at fig. 3.10 we see that the coordinate of the center x_0, y_0 can be obtained by finding the four points S, R, P and Q. By definition, the crossing of the two segments $\overline{SR}$ and $\overline{PQ}$ uniquely identifies the center of the ellipse. The points S and R are obtained by imposing $y = y_0$ in equation 3.37 while the points P and Q are obtained by imposing $x = x_0$ in the same equation. We have:

$$\begin{cases} Ax^2 + 2Bxy + Cy^2 + 2Dx2 + Ey + f = 0 \\ y = y_0 \end{cases} \tag{3.47}$$

Operating the substitution, eq. 3.47 becomes:

$$Ax^2 + 2Bxy_0 + Cy_0^2 + 2Dx + 2Ey_0 + F = 0$$
$$Ax^2 + (2By_0 + 2D)x + Cy_0^2 + (2Ey_0 + F) = 0 \quad (3.48)$$

Eq. 3.48 is a second-degree equation with two solutions x_1 and x_2. The x-coordinate of the center is obtained by noticing that $x_0 = (x_1 + x_2)/2$. If we take the average of the two solutions of a second-degree algebraic equation $ax^2 + bx + c = 0$, it is easy to see that the average is $-\frac{b}{a}$. We therefore have:

$$x_0 = -\frac{1}{2}\frac{b}{a} = -\frac{(By_0 + D)}{A} \quad (3.49)$$

where a and b are respectively the coefficients of the quadratic x^2 and linear x terms in eq. 3.48. Similarly, we find that the coordinate y_0 of the center of the ellipse is:

$$y_0 = -\frac{(Bx_0 + E)}{C} \quad (3.50)$$

Eqs. 3.49 and 3.50 can be rewritten as:

$$\begin{cases} Ax_0 + By_0 + D = 0 \\ Cy_0 + Bx_0 + E = 0 \end{cases} \quad (3.51)$$

and now you see why we used the factors 2. After a little bit of algebra, the coordinates of the center of the ellipse are given by:

$$x_0 = -\frac{(BE - CD)}{B^2 - AC}, \quad y_0 = -\frac{(AE - BD)}{B^2 - AC} \quad (3.52)$$

Having found the coordinates of the center of the ellipse , a coordinate transformation $x \to x + x_0$ and $y \to y + y_0$ will shift the center of the ellipse at the center of the new coordinate system. Let's verify that this is indeed the case. With the substitution $x \to x + x_0$ and $y \to y + y_0$, eq. 3.37 becomes:

$$A(x+x_0)^2 + 2B(x+x_0)(y+y_0) + Cy^2 + 2D(x+x_0) + 2E(y+y_0) + F = 0 \quad (3.53)$$

If we collect the various terms, we have:

$$Ax^2 + Ax_0^2 + 2Axx_0 + Cy^2 + Cy_0^2+$$
$$+2Cyy_0 + 2Bxy + 2Bxy_0 + 2Bx_0y + 2Bx_0y_0+ \quad (3.54)$$
$$+2Dx + 2Dx_0 + 2Ey + 2Ey_0 + F = 0$$

We can re-order the terms of eq. 3.54 to have:

$$
\begin{aligned}
&Ax^2 + 2Bxy + Cy^2 + \\
&+2x(Ax_0 + By_0 + D) + \\
&+2y(Bx_0 + Cy_0 + E) + \\
&+x_0(Ax_0 + By_0 + D) + \\
&+y_0(Bx_0 + Cy_0 + E) + \\
&+Ey_0 + Dx_0 + F = 0
\end{aligned}
\tag{3.55}
$$

The terms in parenthesis in the 2nd, 3rd, 4th and 5th row of eq. 3.55 are all zero because of eq. 3.51. We have, therefore:

$$
Ax^2 + 2Bxy + Cy^2 + (Ey_0 + Dx_0 + F) = 0 \tag{3.56}
$$

where we managed to eliminate the linear terms in x and y. Now the conic (with the exception of the parabola) is centered at the origin of our coordinate system x, y. Notice that the constant term has now changed from F to $(Ey_0 + Dx_0 + F)$. Eq. 3.56 can be written in matrix form as:

$$
(x \ y) \begin{pmatrix} A & B \\ B & C \end{pmatrix} \begin{pmatrix} x \\ y \end{pmatrix} + (Ey_0 + Dx_0 + F) = 0 \tag{3.57}
$$

We are left with the last step, i.e. a rotation that will bring the axes of the ellipse to be parallel to the new coordinate system. We need a transformation where the new axes are rotated to be exactly aligned with the axis of the conic section. We have already stated that a coordinate transformation is represented by a matrix and therefore we need to find a matrix (or a group of matrices) that does align the axis.

In order to find this matrix, we notice that if we find an operation that makes the 2×2 matrix of eq. 3.57 diagonal, then the axis will be aligned. This process is called *diagonalization* of a matrix and is a very powerful tool in many physical situations.

Definition: A vector $\vec{u}$ and a scalar λ such that $A\vec{u} = \lambda\vec{u}$ are called respectively the **eigenvector** and **eigenvalue** of A.

The eigenvalues are found by solving the so-called *characteristic equation* associated with the matrix Q:

$$
det(Q - \lambda I) = det \begin{pmatrix} A - \lambda & B \\ B & C - \lambda \end{pmatrix} = (A - \lambda)(C - \lambda) - B^2 = 0 \tag{3.58}
$$

The characteristic equation 3.58 becomes:

$$
(A - \lambda)(C - \lambda) - B^2 = \lambda^2 - \lambda(A + C) + (AC - B^2) = 0 \tag{3.59}
$$

which has the two solutions[8] (eigenvalues) λ_1 and λ_2 given by:

$$\lambda_1 = \frac{-(A+C) + \sqrt{(A-C)^2 - 4B^2}}{2}$$
$$\lambda_2 = \frac{-(A+C) - \sqrt{(A-C)^2 - 4B^2}}{2} \tag{3.60}$$

In order to find the eigenvectors , we need to solve the two equations:

$$(Q - \lambda_1 I)\vec{u} = 0$$
$$(Q - \lambda_2 I)\vec{v} = 0 \tag{3.61}$$

Solving eq. 3.61 will produce the components of two eigenvectors for which we are given only the direction. By normalizing them to one, i.e. $\vec{e}_u = \frac{\vec{u}}{\|u\|}$ and $\vec{e}_v = \frac{\vec{v}}{\|v\|}$, we get the unit vectors of the rotated coordinate system where the conic section has been aligned.

Having calculated the eigenvalues λ_1 and λ_2, the new representation of the rotated conic section is given by:

$$(x \ \ y) \begin{pmatrix} \lambda_1 & 0 \\ 0 & \lambda_2 \end{pmatrix} \begin{pmatrix} x \\ y \end{pmatrix} + (Ey_0 + Dx_0 + F) = 0 \tag{3.62}$$

With a bit of extra work we can find a simpler expression for the constant term $(Ey_0 + Dx_0 + F)$. Let's recall the form of the matrix A_Q from eq. 3.45. Its determinant is:

$$\det \begin{pmatrix} A & B & D \\ B & C & E \\ D & E & F \end{pmatrix} = A(CF - E^2) - B(BF - ED) + D(BE - CD) \tag{3.63}$$

Another useful determinant is $\det(Q) = (AC - B^2)$. With a bit of algebra, it can be shown that eq. 3.62 can be written as:

$$(x \ \ y) \begin{pmatrix} \lambda_1 & 0 \\ 0 & \lambda_2 \end{pmatrix} \begin{pmatrix} x \\ y \end{pmatrix} + \frac{\det(A_Q)}{\det(Q)} = 0 \tag{3.64}$$

If we now use the property that determinants of matrices are invariant , then $\det(Q) = \lambda_1\lambda_2$ and equation 3.64 becomes:

$$(x \ \ y) \begin{pmatrix} \lambda_1 & 0 \\ 0 & \lambda_2 \end{pmatrix} \begin{pmatrix} x \\ y \end{pmatrix} + \frac{\det(A_Q)}{\lambda_1\lambda_2} = 0 \tag{3.65}$$

We conclude this (long) excursion by writing eq. 3.65 in algebraic form:

---

[8]It can be proved, and we do not do it here, that the eigenvalues of a symmetric matrix are real.

$$\lambda_1 x^2 + \lambda_2 y^2 = -\frac{det(A_Q)}{\lambda_1 \lambda_2}$$

$$\frac{x^2}{\lambda_2} + \frac{y^2}{\lambda_1} = -\frac{det(A_Q)}{\lambda_1^2 \lambda_2^2}$$

(3.66)

In summary, given a conic section of the general Cartesian form given by eq. 3.37, it is possible to translate and rotate the conic such that it is centered at the origin of the Cartesian coordinate system and with axes aligned with the x and y axis. We say that the conic section is in **canonical form**.

Here we report the canonical forms of the conic sections:

1. circle: $x^2 + y^2 = r^2$

2. ellipse: $\frac{x^2}{a^2} + \frac{y^2}{b^2} = 1$

3. parabola: $y^2 = 4ax$

4. hyperbola: $\frac{x^2}{a^2} - \frac{y^2}{b^2} = 1$

Notice that none of the canonical forms (except the parabola) have linear terms in x and y, nor mixed terms in xy.

Let us now study in detail the canonical form of the ellipse (see fig. 3.11). Let us first verify that the ellipse in fig. 3.11 is described by the canonical form. The formula expressing the distance between two points (x_1, y_1) and (x_2, y_2) in the Cartesian plane is:

$$d = \sqrt{(x_2 - x_1)^2 + (y_2 - y_1)^2}$$

(3.67)

Eq. 3.67 is simply the Pythagorean Theorem.

An alternative definition of the ellipse is the following: the ellipse is the locus of points $P(x, y)$ for which the distance $\overrightarrow{F_1 P} + \overrightarrow{PF_2} = 2a$ is constant. By placing the point P coincident with the point $P_1(a, 0)$ we immediately verify that the constant is $= 2a$ where a is the semi-major axis. Let us now verify that the condition $\overrightarrow{F_1 P} + \overrightarrow{PF_2} = 2a$ gives the ellipse in canonical form.

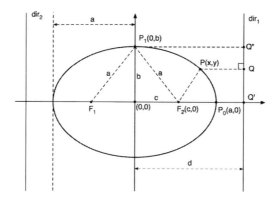

FIGURE 3.11 Canonical ellipse centered at the origin and with foci F_1 and F_2 along the x axis.

$$\overrightarrow{F_1 P} + \overrightarrow{PF_2} = 2a$$

$$\sqrt{(x+c)^2 + y^2} + \sqrt{(x-c)^2 + y^2} = 2a$$

$$\sqrt{(x+c)^2 + y^2} = 2a - \sqrt{(x-c)^2 + y^2}$$

$$(x+c)^2 + y^2 = 4a^2 + (x-c)^2 + y^2 - 4a\sqrt{(x-c)^2 + y^2}$$

$$x^2 + c^2 + 2xc + y^2 = 4a^2 - 4a\sqrt{(x-c)^2 + y^2} + x^2 + c^2$$
$$- 2xc + y^2$$

$$2xc = 4a^2\sqrt{(x-c)^2 + y^2} - 2xc$$

$$a^2 - a\sqrt{(x-c)^2 + y^2} - xc = 0$$

$$a\sqrt{(x-c)^2 + y^2} = a^2 - xc$$

$$a^2[(x-c)^2 + y^2] = a^4 + x^2c^2 - 2a^2xc$$

$$a^2x^2 + a^2c^2 + a^2y^2 = a^4 + x^2c^2$$

$$x^2(a^2 - c^2) + a^2y^2 = a^2(a^2 - c^2)$$

$$x^2b^2 + a^2y^2 = a^2b^2$$

$$\frac{x^2}{a^2} + \frac{y^2}{b^2} = 1$$

$$(3.68)$$

where b is called the *semi-minor* axis.

In fig. 3.11 we see an ellipse centered at the origin $(0,0)$ with foci F_1 and F_2 along the x axis. The two axes $dir1$ and $dir2$ are called *directrix* and define the ellipse through the relationship

$$e = \frac{\overline{PF_2}}{\overline{PQ}} \tag{3.69}$$

where the constant e is called the **eccentricity**. If the point P is coincident with the point P_1 then again using the Pythagorean theorem, we see that $b^2 + c^2 = a^2$.

Let us now calculate the distance from the directrix $dir1$ to the origin $\overline{OQ'}$. If we place the point P coincident with Q', we can write eq. 3.69 as:

$$e = \frac{\overline{P_0 F_2}}{\overline{P_0 Q'}} = \frac{(a-c)}{d-a} \tag{3.70}$$

Equivalently, if we place the point P coincident with P_1, we can write:

$$e = \frac{\overline{P_1 F_2}}{\overline{P_1 Q''}} = \frac{a}{d} \tag{3.71}$$

Equating 3.70 and 3.71 we have:

$$e = \frac{(a-c)}{d-a} = \frac{a}{d} \tag{3.72}$$

Solving for d gives:

$$d = \frac{a^2}{c}, \quad e = \frac{c}{a} \tag{3.73}$$

The **focal parameter** $p = \frac{a(1-e^2)}{e}$ is the distance between the focus and its closest directrix and is equal to the segment $\overline{F_2 Q'}$. We have:

$$p = \overline{F_2 Q'} = \overline{P_0 F_2} + \overline{P_0 Q'}$$
$$\overline{P_0 F_2} = a - c$$
$$\overline{P_0 Q'} = d - a = \frac{a^2}{c} - a$$
$$p = \frac{a^2}{c} - a + a - c = \frac{a^2}{c} - c = \frac{a^2 - c^2}{c} \tag{3.74}$$
$$c = ae$$
$$p = \frac{a^2 - c^2}{c} = \frac{a^2 - a^2 e^2}{ae}$$
$$p = \frac{a(1 - e^2)}{e}$$

We conclude this section by showing the equation of a conic section in polar coordinates. In fig. 3.12 the geometry of the polar definition of a conic section is shown. We have already seen that the definition of a conic section requires that the $P(x, y)$ of the conic is such that the distance $\overline{PF}$ to the fixed focus F

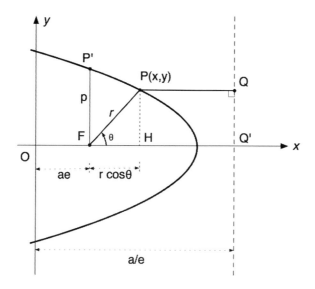

FIGURE 3.12 Geometry to express conic sections in polar form.

is proportional to the distance to the directrix line $\overline{PQ}$, i.e. $\overline{PF} = e\overline{PQ}$ where the constant e is the eccentricity. From fig. 3.12 we can immediately write the conic condition:

$$\overline{PF} = e\overline{PQ}$$
$$r = \left(\frac{a}{e} - ae - r\cos\theta\right) \tag{3.75}$$

which simplifies to:

$$r = \frac{a(1 - e^2)}{(1 + e\cos\theta)} = \frac{p}{(1 + e\cos\theta)} \tag{3.76}$$

The above equation lets us express the **semi-latus rectum**[9] $p = \overline{FP'}$:

$$p = a(1 - e^2) \tag{3.77}$$

3.2.2 Kepler's 1st Law as Discovered by Kepler Himself in the Years 1600 - 1630

At the time of Kepler, the regularity of the motion of stars and planets was assumed to be a reflection of the perfection of the Creator. Therefore only

---

[9]The latus rectum of a conic section is the chord through a focus parallel to the directrix of the conic section.

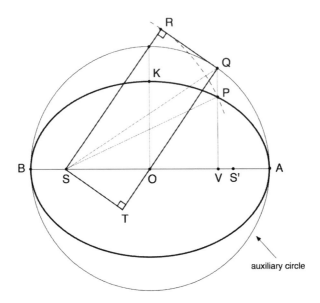

FIGURE 3.13 Kepler's construction of an elliptical orbit.

perfect geometrical figures were considered, like the circle. However, Kepler was quite modern in his quest to explain the available observations of the motion of planets in the sky with respect to the *fixed* stars through the usage of mathematical reasoning. He believed that the mechanisms of nature can be understood and the explanation would contain some inner beauty. This view is shared still today by (probably) the majority of scientists. Kepler was a talented mathematician and was lucky enough to attract the attention of Tycho Brahe, who collected 30 years of accurate measurements of positions of planets in the sky.

Kepler happened to be the right person at the right time, with the proper attitude toward detailed calculations and accuracy. Kepler was after an explanation of the motion of planets and he was looking for any law that would help him understand the large amount of data he had access to. In addition, he adopted the Copernican view that the Sun was at the center of the orbits of the planets and this view was certainly quite novel for his times. It is probable that Kepler liked the Copernican system because it was simpler than the alternative proposal of having a geocentric system requiring complex superposition of circumferences whose center was moving along other circumferences (epicycles).

In line with his thinking that the motion of sky objects must be obtained with the aid of circles, Kepler started to construct complex geometries to try to find the best match to Tycho's data. In this approach Kepler showed how modern his thinking was: he constructed a theory based on geometric intuition, then he checked that the theory correctly explained the data. If it

FIGURE 3.14 Tycho Brahe was a very accurate Danish astronomer. He generated a large amount of good-quality observational data used later by Kepler.

Nikolaus Kopernikus.

FIGURE 3.15 Nicolaus Copernicus was a Polish mathematician who formulated the first successful theory of heliocentrism.

did not, then he kept searching. He eventually found a geometric construction (see fig. 3.13) that was capable of correctly validating the data he had for the motion of Mars. It is important to underline that Kepler was looking for a **physical theory** to be validated by experimental data and not just finding the best curve approximating the planetary motions (see [2]).

Looking at fig. 3.13, let us draw the main auxiliary circle[10] of center O and diameter $\overline{AB}$. Let us choose an arbitrary point Q on the auxiliary circle and draw a segment $\overline{QV}$ perpendicular to the diameter $\overline{AB}$. From Q again, draw a straight line to O extending it to the point T such that the segment $\overline{TS}$ is perpendicular to $\overline{QT}$. This identifies the point S, which we will see later is one of the two foci of the constructed ellipse. Take a drawing compass pointed at S and select a length $\overline{SR}$ equal to the segment $\overline{QT}$. Draw a circle centered at S from R to intercept the segment $\overline{QV}$ at the point P. Choosing other points like Q on the main auxiliary circle and repeating the same procedure will obtain the ellipse with semi-minor axis $b = \overline{OK}$ and semi-major axis $a = \overline{OA}$. It can be shown [5] that this construction is such that:

$$\frac{\overline{PV}}{\overline{QV}} = \frac{\overline{OK}}{\overline{OA}} = \frac{b}{a} \qquad (3.78)$$

and it is an alternative definition of the ellipse known since Apollonius of Perga around 200 B.C. [3].

So Kepler identified correctly the shape of the orbit of the planet Mars. Unfortunately he did not have the physics to explain why the orbit had that specific shape. We now turn our attention to the details of this problem and we will solve it using different techniques. Some will appear a bit convoluted and others will show a beautiful simplicity but all are useful and will allow us to look at this central problem from various perspectives.

3.2.3   Kepler's Problem: Geometrical Solution

The Kepler problem consists in the determination of the functional form with the distance of the force that is exerted between the Sun and a planet such that the orbit is elliptical. The inverse Kepler problem is the opposite: given a force, what is the shape of the resulting orbit. We will now proceed to both proofs using only geometrical arguments. In certain cases we will use small intervals of time Δt, space ΔR, or velocity Δv, and narrow triangles or segments. This is not an attempt to hide a limiting process and therefore the use of calculus. The main difference between the proof we now give and the proof using calculus is the fact that the geometrical proofs are somehow *more general* and do not require the transition to infinitesimal quantities and the limit operations that we have encountered earlier in the book. The Δ-quantities are not required to go to zero. That is the reason why, in few places

---

[10]There is also a minor auxiliary circle of radius $\overline{OK}$ with center O and tangent to the ellipse in K and the symmetric point to K not shown.

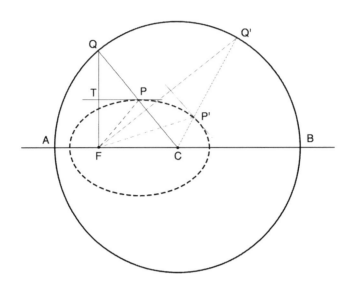

FIGURE 3.16 Another geometric construction of the ellipse.

in the following, we need to be a bit convoluted when introducing auxiliary geometrical construction. The theory of orbits with calculus will be done later once we have the solid knowledge of the geometrical theory.

There is a story regarding Edmund Halley (the discoverer of the famous comet) and Isaac Newton about the inverse Kepler problem. In 1684 Halley had the curiosity to meet the famous Newton in Cambridge to discuss the mechanics of celestial bodies. At that time there were a lot of discussions about the source of the force that is capable of keeping the planets on their orbits and there was the suggestion that a force proportional to the inverse square of the distance might have been the right answer but nobody was able to demonstrate this conjecture. It was already known, thanks to Kepler, that the orbits were elliptical and Halley wanted to see if Newton knew how to demonstrate the conjecture. After all, he (later) invented differential calculus and proposed that the gravitational attraction between two bodies was proportional to the product of their masses and inversely proportional to the square of the distance between them, so it seemed natural for Halley to ask Newton. When asked on the spot, Newton claimed that he had calculated it but somehow he misplaced it and asked Halley to be patient. Newton effectively sent a demonstration to Halley a few months later but he did not want to publish it because he was not satisfied. So Newton kept working on it until few years later he published not only the demonstration but much more: he published the *Principia*. We probably need to thank Halley for triggering such monumental work!

Let us state again what we will be demonstrating: we start from Kepler's equal area in equal times law (Kepler 2nd) and by using Newton's 1st law we

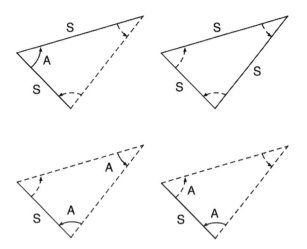

FIGURE 3.17 Congruence of two triangles (SAS - Side-Angle-Side), (SSS - Side-Side-Side), (ASA - Angle-Side-Angle) and (AAS - Angle-Angle-Side).

deduce that the source of the force attracting the planets is directed along the line connecting the planet and the Sun. We then use Kepler's 3rd law, orbital periods are proportional to distance from the Sun to the power of $3/2$, together with Newton's laws to show that the force of gravity is proportional to the inverse square of the distance between the planet and the Sun. Finally, again using Newton's laws, we will deduce that the orbit is elliptical[11].

Let us discuss another interesting geometrical fact about circles and ellipses: we will now show that given a circle and a point off-center located inside it, it is possible to geometrically build an ellipse.

With reference to fig. 3.16, we start by arbitrarily choosing two points F and C. Now we draw from the point F a line in any direction we like $\overline{FQ}$ (another choice could be $\overline{FQ'}$, which will produce a different point of the ellipse). Let's choose a point T over the line $\overline{FQ}$ through which we trace a perpendicular line $\overline{TP}$ which bisects the segment $\overline{FQ}$, i.e. cuts it in half so that $\overline{FT} = \overline{TQ}$. The line $\overline{TP}$ has two important properties: first, if we draw the line $\overline{CQ}$, then we have that $\overline{PF}\overline{PQ}$ and so the point P belongs to the ellipse shown with a dashed line in fig. 3.16; second, the line $\overline{TP}$ identifies the **tangent** to the ellipse at the point P. In fact, as the point Q moves around the circle, called the **directrix circle**, (for example, to point Q'), the

---

[11]More accurately, we will show that the orbit is a conic section.

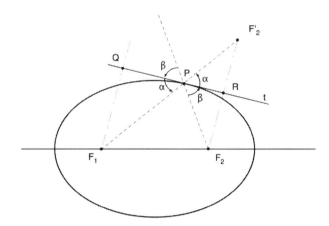

FIGURE 3.18 The tangent t at P is the external angle bisector of the angle $\angle F_1PF_2$, i.e. $\alpha = \beta$.

corresponding points obtained with the recipe above trace the ellipse. In other words, there is a one-to-one correspondence between the ellipse and the circle.

We prove the first property if we show that the two triangles $\triangle FTP$ and $\triangle QTP$ are congruent[12]. Two triangles are congruent (see fig. 3.17) if:

1. SAS (Side-Angle-Side): two pairs of sides of two triangles are equal in length, and the angles in between them are equal;

2. SSS (Side-Side-Side): three pairs of sides of two triangles are all equal in length;

3. ASA (Angle-Side-Angle): two pairs of angles of two triangles are equal, and the sides between them are equal in length;

4. AAS (Angle-Angle-Side): two pairs of angles of two triangles are equal, and a pair of corresponding non-included sides are equal.

The two triangles $\triangle FTP$ and $\triangle QTP$ are congruent because of SAS: $\overline{PT}$ is shared, $\overline{QT} = \overline{TF}$ by construction, and angles $\angle QTP = \angle FTP$ are both right angles by construction. If the two triangles are congruent, then $\overline{PQ} = \overline{CP}$. But eq. 3.68 tells us that an alternative definition of an ellipse is by requiring that the distance $\overline{PQ} + \overline{CP} = $ constant. We know a bit more: from eq. 3.68 we know the value of the constant to be twice the semi-major axis of the ellipse, $\overline{PQ} + \overline{CP} = 2a$. Therefore the big circle centered at C also has radius equal to $2a$.

We now prove the second property, i.e. that the line $\overline{TP}$ in fig. 3.16 identifies the tangent to the ellipse at the point P. We already have defined the

---

[12]Let's repeat here that congruent means exactly equal in size and shape.

tangent line to a curve when we discussed instantaneous velocity (see eqs. 1.24 and 1.25). Here we prefer to use a different equivalent definition: a tangent to a curve at a point P is a line that has only one point P in common with the curve.

We first prove that the tangent line is the external bisector[13] of the angle $\angle F_1 P F_2$ in figs. 3.18 and 3.19. In fig. 3.18 the two lines from F_1 and F_2 to the line t are traced to be perpendicular to t. These two lines generate two triangles $\triangle F_1 PQ$ and $\triangle F_2 PR$. Let's call the segment $m = \overline{PQ}$ and $k = \overline{RQ}$. Applying the Pythagorean Theorem we have:

$$
\begin{aligned}
\overline{F_1 P}^2 &= m^2 + \overline{F_1 Q}^2 \\
\overline{PF_2}^2 &= (k - m)^2 + \overline{F_2 R}^2
\end{aligned}
\tag{3.79}
$$

Now we ask: what is the value of m that minimizes the path $y = F_1 P F_2$? This is equivalent to asking that the derivative with respect to m of the path length y be zero:

$$
\frac{d}{dm}(\overline{F_1 P} + \overline{PF_2}) = \frac{dy}{dm} = 0
$$
$$
\frac{d}{dm}(\sqrt{m^2 + \overline{F_1 Q}^2} + \sqrt{(k - m)^2 + \overline{F_2 R}^2}) = 0
\tag{3.80}
$$

remembering that $\frac{d}{dx}\sqrt{x} = \frac{1}{2\sqrt{x}}$, we have:

$$
\frac{dy}{dm} = \frac{m}{\sqrt{m^2 + \overline{F_1 Q}^2}} - \frac{(k - m)}{\sqrt{(k - m)^2 + \overline{F_2 R}^2}} = 0
\tag{3.81}
$$

the two terms in eq. 3.81 are respectively $\sin\alpha$ and $\sin\beta$. Therefore the minimal path is obtained when $\sin\alpha = \sin\beta \rightarrow \alpha = \beta$. The minimal path is also the path that a light ray will follow according to Fermat's principle of least time. If a light ray originated in F_1, it will hit the ellipse in P with a certain angle α. If we assume that the inside of the ellipse is made of a reflecting material, like a mirror for example, then the light ray is reflected exactly towards F_2 with the same angle α. Or, in optics terms, the incidence angle is equal to the reflection angle. This means that if we rotate an ellipse around its major axis, we obtain an ellipsoid shape which, if internally reflecting, has the beautiful property that any light ray originated in F_1 will end up in F_2. Laser cavities or even pizza ovens are designed based on this focusing property of the ellipse. This property is the reason why the two special points F_1 and F_2 are called **foci**.

Now we are ready to prove that the line t is tangent to point P. With

---

[13]The external angle bisectors of a triangle $\triangle ABC$ are the lines bisecting the angles formed by the sides of the triangles and their extensions.

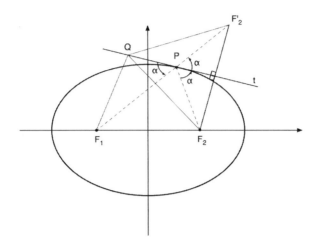

FIGURE 3.19 Geometry to show that the line t is tangent at point P of the ellipse.

reference to fig. 3.19, consider a point Q different from P on the line t exterior angle bisector and let F_2' be the point you get by reflecting F_2 across the line t. We have that $\overline{QF_2} = \overline{QF_2'}$ and $\overline{PF_2} = \overline{PF_2'}$.

Since $\overline{PQ} \equiv t$ is the exterior angle bisector of $\angle F_1 P F_2$, the point P lies on the line $\overline{F_1 F}$. Now let's look at the straight line $\overline{F_1 F_2'}$ through P and compare it with the path $F_1 Q F_2' = \overline{F_1 Q} + \overline{Q F_2'}$. Clearly the path $\overline{F_1 Q} + \overline{Q F_2'}$ must be longer than the path $\overline{F_1 F_2'}$ because the shortest path through two points in Euclidean geometry is the straight line. It follows that the point Q cannot be on the ellipse because $F_1 Q F_2 > F_1 P F_2$. So the *only* point on the exterior angle bisector belonging to the ellipse must be P and therefore t is tangent in P.

At this point we have reviewed all the math needed to prove Kepler problems. We now turn to the physics, specifically Newton's laws, and prove Kepler's second law of equal areas swept in equal times. Newton's first law (inertia) tells us that a planet with no external forces applied, either stays still or moves at constant velocity (remember that velocity is a vector and therefore constant velocity means constant speed and constant direction).

Let's first prove that the equal areas law is valid in the special case of absence of forces. In fig. 3.20 a planet is moving with respect to the Sun S from A to E at constant velocity $v = \frac{\overline{AB}}{\Delta t} = \frac{\overline{BC}}{\Delta t} = \frac{\overline{CD}}{\Delta t} = ...$ We are assuming that the Sun does not exert any gravitational force on the body. It is easy to see that all the triangles 1, 2, 3, etc. all have the same area because they have equal base and they all share the same height $\overline{SA}$.

Let us now "switch on" the gravitational attraction from the Sun. This means that, because of Newton's second law, the planet now will feel a force

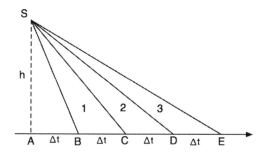

FIGURE 3.20 Equal areas law in absence of external forces.

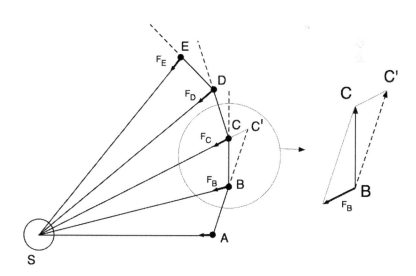

FIGURE 3.21 Newton's geometrical construction of the orbital path of a planet under a central force directed towards the Sun.

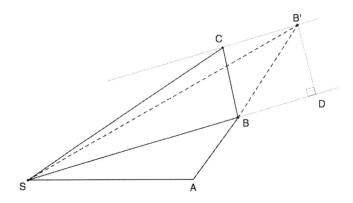

FIGURE 3.22 Geometry to prove the equal areas law in an orbit with central force.

directed exactly towards the Sun (see fig. 3.21). Newton imagined that the planet moves from A to B with constant velocity and then receives an instantaneous impulse of force F_B when in B. Because of Newton's second law, the force F_B is directed exactly towards the Sun. The effect is that the planet, instead of progressing freely to C', bent to C. If the vector indicated by F_B has a length equal to the amount the force would displace if the planet did not have lateral motion, then the actual travel of the planet will be the parallelogram between the displacement vectors $\overline{BC'}$ and F_B or, in vector notation, $\overrightarrow{BC} = \overrightarrow{F_B} + \overrightarrow{BC'}$. The real trajectory will be obtained by making the segments infinitesimal so that the orbit becomes a continuous curve. Note that the segment $\overline{CC'}$ is parallel to the vector F_B by construction.

We now prove that Kepler's area law is valid also in the case of a planet orbiting the Sun. To this end we enlarge a portion of fig. 3.21 into fig. 3.22.

We want to prove that the area of the triangle $\triangle SAB$ is equal to the area of the triangle $\triangle SBC$. We just proved that the area of $\triangle SAB$ is equal to the area of $\triangle SBB'$, which is the case of absence of external forces on the planet. Now if we prove that the area of $\triangle SBB'$ is equal to the area of $\triangle SBC$, we are done. This is easily demonstrated by noticing that the line $\overline{CB'}$ is parallel to the line $\overline{SB}$ and therefore the two triangles $\triangle SBB'$ and $\triangle SBC$ share the same base $\overline{SB}$ and the same height $\overline{CB} = \overline{B'D}$. We just proved that Kepler's second law is a consequence of Newton's first and second laws. Note that we only assumed that the force is *central*, i.e. directed towards the Sun, and so the inverse square law with distance dependence of the force has no connection with Kepler's second law.

Let us discuss the value of the constant in the Kepler's second law using a bit of basic calculus. The area swept by the radius vector $\overrightarrow{r}(t)$ in the small time interval Δt is indicated by ΔA in fig. 3.23 and is equal to the area of the

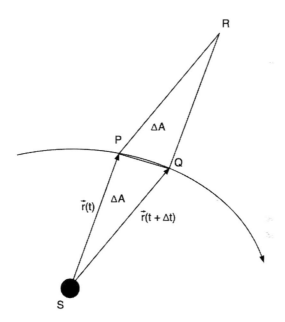

FIGURE 3.23 Geometry to show that area swept in unit time is proportional to angular momentum.

triangle $\triangle SPQ$. We already have seen (see fig. 2.7) that the vector product represents the area of the parallelogram identified by the two vectors being multiplied. We have that the area ΔA is half the area of the parallelogram $SPQR$. Therefore we have:

$$\Delta A = \frac{1}{2}\vec{r}(t) \times \vec{r}(t + \Delta t) \tag{3.82}$$

The area swept in unit time is then:

$$\begin{aligned}\frac{\Delta A}{\Delta t} &= \frac{1}{2\Delta t}\vec{r}(t) \times \vec{r}(t + \Delta t) \\ &= \frac{\vec{r}(t) \times [\vec{r}(t) + \dot{\vec{r}}(t)\Delta t]}{2\Delta t}\end{aligned} \tag{3.83}$$

because of the sine term, the vector product of a vector with itself is zero. The time derivative of the vector $\vec{r}$ is the velocity vector $\vec{v}$. We therefore have:

$$\begin{aligned}\frac{\Delta A}{\Delta t} &= \frac{\vec{r}(t) \times \dot{\vec{r}}(t)\Delta t}{2\Delta t} \\ &= \frac{1}{2}\vec{r} \times \vec{v} \\ &= \frac{\vec{L}}{2m}\end{aligned} \tag{3.84}$$

where we define the angular momentum vector as:

$$\begin{aligned}\vec{L} &= \vec{r} \times m\vec{v} \\ &= m\vec{r} \times (\vec{\omega} \times \vec{r}) \\ &= m\vec{\omega}(\vec{r} \cdot \vec{r}) - m\vec{r}(\vec{r} \cdot \vec{\omega}) \\ &= mr^2\vec{\omega} \\ &= mr^2\frac{d\theta}{dt} \\ &= mr^2\dot{\theta}\end{aligned} \tag{3.85}$$

We see that $\vec{L}$ is directed as $\vec{\omega}$, i.e. perpendicular to the plane of the orbit. In eq. 3.85 we have used the triple vector product formula 2.36 and the fact that $\vec{r}$ and $\vec{\omega}$ are perpendicular, and therefore their scalar product is zero. Therefore, Kepler's second law translates into the conservation of angular momentum for a central force.

Proof of the Kepler problem. We are now ready to prove that an inverse square law force directed to the Sun is a consequence of Kepler's third law. We will use the geometrical fact that the length of the circumference of a circle of radius x is $L = 2\pi x$. Let us restrict ourselves to the case of a circular orbit. The speed of a planet in circular orbit is constant while the direction changes. The constant speed can be written as the length of the circumference divided by the time needed for the planet to complete one revolution. This time is called the **period** and is indicated with T. We have that the speed is:

$$v = \frac{2\pi R}{T}$$

where R is the radius of the circular orbit. The speed is the total amount of space traveled by the planet divided by the period. Let us now look at the changes of the velocity: the velocity vector keeps its magnitude constant but changes direction continuously. If the velocity vector starts at a time $t = 0$ in a certain direction, it will return to that point exactly in the same direction after a full revolution of the planet, i.e. after exactly one period T. In analogy with eq. 3.2.3 which gives the total amount of space in one period, we can write that the total amount of velocity change[14] in one period is:

$$\frac{\Delta v}{\Delta t} = \frac{2\pi v}{T}$$

Newton's second law (see eq. 2.57) tells us that the force directed towards the Sun is proportional to the change in velocity. Therefore we have:

$$F \propto \frac{\Delta v}{\Delta t} = \frac{2\pi v}{T} = \frac{(2\pi)^2 R}{T^2} \qquad (3.86)$$

Kepler's third law tells us that $T^2 \propto R^3$. So, eq. 3.86 becomes:

$$F \propto \frac{(2\pi)^2 R}{T^2} = \frac{(2\pi)^2 R}{R^3} \rightarrow F \propto \frac{1}{R^2} \qquad (3.87)$$

We just proved geometrically that Kepler's equal area law together with Newton's first and second laws imply that the central force must be an inverse square law of the distance between the Sun and the planet.

Proof of the Kepler inverse problem. We need to prepare the proof with some more geometrical facts. Following Feynman[15] [7] we proceed to prove that the magnitude of the variation of velocity of a planet orbiting the Sun is constant and therefore will only change direction in such a way that it is constantly pointing towards the Sun (Newton's second law). We can represent Kepler's equal areas law in fig. 3.24 (a) where the two shaded areas are swept in equal time Δt. It follows that the time is proportional to the area swept, $\Delta t \propto$ (area swept).

We now proceed to prove that the area swept is proportional to the distance

---

[14]The amount of velocity change in one period is the average acceleration which for a uniform circular motion is constant.

[15]Richard P. Feynman was one of the most influential physicists of the 20th century. He won the Nobel prize in 1965 (together with S. Tomonaga and J. Schwinger) for his work on quantum electrodynamics.

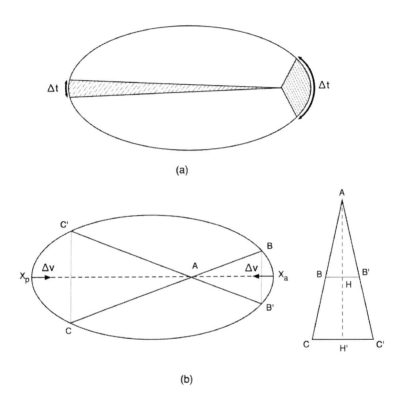

(a)

(b)

FIGURE 3.24 (a) Kepler's first law of equal areas in equal times. The two shaded areas are equal. (b) Geometry to show that area is proportional to R^2. The portion of orbit CC' sweeps the same angle of the portion of orbit BB'.

FIGURE 3.25 Richard P. Feynman was an American theoretical physicist who made substantial contributions to quantum mechanics and
quantum electrodynamics.

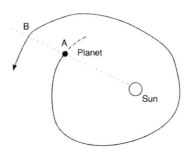

FIGURE 3.26 A planet crossing a half-line from the Sun twice at points A and B. It can be shown that in an inverse square central force, points A and B must be coincident and the curve is closed.

to the Sun squared (area swept) $\propto R^2$. Fig. 3.24 (b) on the left shows two portions of orbit corresponding to the same angle, since the angle $\angle CAC'$ is opposite to angle $\angle BAB'$. Without losing generality we can approximate the curved segments CC' and BB' with their straight segments $\overline{CC'}$ and $\overline{BB'}$. Let us now consider the triangles on the right of fig. 3.24 (b) obtained by rotating the triangle $\triangle ABB'$ around the point A by $180°$. The two triangles $\triangle ABB'$ and $\triangle ACC'$ are similar by construction. We have:

$$\frac{\overline{AH}}{\overline{BB'}} = \frac{\overline{AH'}}{\overline{CC'}} \tag{3.88}$$

and the area of the triangle $\triangle ABB'$ is $\frac{1}{2}\overline{AH}\cdot\overline{BB'}$. Now we study what happens to this area when $\overline{AH'} = 2\overline{AH}$, i.e. we double the distance to the Sun. Eq. 3.88 becomes:

$$\frac{\overline{AH}}{\overline{BB'}} = \frac{2\overline{AH}}{\overline{CC'}} \tag{3.89}$$
$$\rightarrow \overline{CC'} = 2\overline{BB'}$$

Doubling the distance to the Sun has also doubled the base of the triangle $\triangle ABB'$. This means that the area of the triangle $\triangle ABB'$ becomes:

$$\frac{1}{2}\overline{AH}\ \overline{BB'} \rightarrow \frac{1}{2}\left(2\overline{AH}\right)\left(2\overline{BB'}\right) = 4\frac{1}{2}\overline{AH}\ \overline{BB'} \tag{3.90}$$

Therefore doubling the distance makes the area $2^2 = 4$ times the area bigger, i.e. (area swept) $\propto R^2$. But we have demonstrated above that $\Delta t \propto$ (area swept) and therefore $\Delta t \propto R^2$.

An inverse square law central force has as a consequence that time inter-

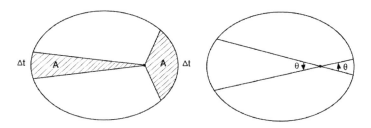

FIGURE 3.27 Equal times equal areas versus equal angle diagrams.

vals along the orbits are proportional to the distance to the Sun squared[16].
Newton's second law tells us that the force $F = \frac{\Delta v}{\Delta t}$ or $\Delta v = F\Delta t$. We have
that:

$$\Delta v = F\Delta t \propto \frac{1}{R^2} \cdot R^2 = 1 \qquad (3.91)$$

Eq. 3.91 tells us that Kepler's second law plus Newton's second law for a
force $\propto \frac{1}{R^2}$ has the important consequence that the magnitude of the velocity
change Δv **is constant and always directed toward the Sun** or, in other
words, Δv does not depend on R. No matter where on the orbit the planet is,
the Δv change is constant provided that the angle is the same.

Let us now find the value of the constant. We can write Newton's second
law as:

$$\vec{a} = -\frac{GM}{r^2}\vec{e}_r = \frac{\Delta v}{\Delta t} = \frac{\Delta v}{\Delta \theta}\frac{\Delta \theta}{\Delta t} \qquad (3.92)$$

where $\vec{e}_r$ is the unit vector connecting the Sun to the planet and $\vec{a}$ is the
acceleration always towards the Sun. In eq. 3.92 we have used the chain rule.

We have seen that Kepler's second law of equal area implies conservation
of angular momentum. This means that in a central force, the magnitude L
is constant. We have:

$$\frac{\Delta \theta}{\Delta t} = \frac{L}{mr^2} \qquad (3.93)$$

Using eq. 3.92, we have:

$$\frac{\Delta v}{\Delta \theta}\frac{L}{mr^2} = -\frac{GM}{r^2}\vec{e}_r \qquad (3.94)$$

Eq. 3.94 shows that the vector Δv is oriented like $\vec{e}_r$ towards the Sun.
Ignoring the unit vector, we have that eq. 3.94 can be re-written as:

---

[16]This demonstration is not rigorous but for our purposes is good enough. A more rigorous
demonstration can be found in [7].

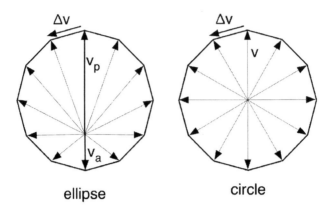

FIGURE 3.28 Hodographs obtained by dividing the orbit of a planet into equal angle sectors. The left panel is relative to an elliptical orbit while the right panel is relative to a circular orbit. Notice that Δv has the same magnitude for each sector. v_p and v_a are respectively the perihelion and aphelion velocities in an elliptical orbit.

$$\|\Delta v\| = \frac{GMm}{L}\Delta\theta \qquad (3.95)$$

Eq. 3.95 tells us that the magnitude of the change in velocity Δv is proportional to the change in angle $\Delta\theta$. If we divide the trajectory of the planet into sectors of equal angles, then the vector sum of all the Δv will make a regular polygon, because the successive changes in the velocity vectors are all inclined to each other at the same angle $\Delta\theta$, and all the magnitudes Δv will be equal. If we imagine taking smaller and smaller $\Delta\theta$, the vector sum of all the magnitudes Δv are equal and will approximate a circle better and better. This circle is called the **hodograph** (see fig. 3.28) and its constant radius is $u = \frac{GMm}{L}$. The important result is that given a central force proportional to the inverse square of the distance, the tips of the velocity vectors around the orbits describe a circle if traced from a common base point. Fig. 3.27, left panel, tells us that the velocity v_p of the planet is maximal at the point closer to the Sun (**perihelion**) and minimal v_a at the opposite point most distant (**aphelion**).

We just proved that an inverse square distance central force generates a circular hodograph. We are left to show that a circular hodograph implies an orbit that is an ellipse. Let us study the relationships between a hodograph and the associated orbit. We already know the two vectors v_a and v_p, correspond to the minimal and maximal orbital speed (see fig. 3.29): their sum is equal to the diameter of the hodograph, i.e. $v_a + v_p = 2u = \frac{2GMm}{L}$. Point Q on the

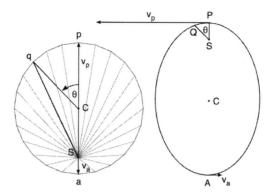

FIGURE 3.29 Relationship between an elliptical orbit and its circular hodograph. The hodograph has been rotated 90° clockwise to better show the two equal angles θ.

orbit, obtained after a rotation of an angle $\theta = \angle QSP$ from the perihelion P, corresponds to the point q on the hodograph where $\theta = \angle qCp$. Similarly the point a on the hodograph (aphelion) corresponds to the point A on the orbit. Notice that in order to have the segments $\overline{Cq}$ and $\overline{SQ}$ be parallel, we needed to rotate the hodograph 90° counterclockwise. Any radius vector for every point on the orbit will correspond to a radius velocity vector on the hodograph displaced by exactly the same angle. Notice also that the velocity at the point Q on the orbit corresponds to the vector $\overrightarrow{Cq}$ on the hodograph. Each angle θ on the hodograph identifies the direction of the tangent to the orbit: evidently this is not enough to build the orbit. The line $\overline{Sq}$ is parallel to the tangent to the point Q while the line $\overline{Cq}$ identifies the direction to the Sun. Clearly the two above conditions do not uniquely identify a point on the orbit. However, we are able to construct an orbit whose shape is the correct one because all the directions are properly identified.

Let's start the construction of the orbit by starting from a given hodograph (see fig. 3.30) of radius $\overline{Cq}$. The line $\overline{Sq}$ is parallel to the velocity at the point Q on the orbit, while the line $\overline{Cq}$ identifies the direction to the Sun. Let's draw the line t obtained by rotating the segment $\overline{Sq}$ around its midpoint. The line t bisects the segment $\overline{Sq}$. The line t intersects the radius $\overline{Cq}$ at a point Q. If we now move the point q around the circle to new points $q_1, q_2, ...$ the new bisecting tangents intersect the corresponding radii at $Q_1, Q_2, ...$ The locus of points Q_i describes an ellipse whose foci are S and C. But we have done exactly the same construction in fig. 3.16, where we demonstrated that such construction generated an ellipse. We have therefore proved geometrically that a circular hodograph is associated with an elliptical orbit. But only an orbit whose central force is proportional to the inverse square of the distance to-

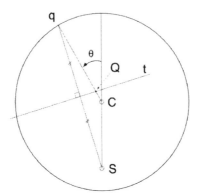

FIGURE 3.30 Construction of an elliptical orbit starting from a circular hodograph and a non-concentric focus S within the circle.

gether with Newton's laws, generates circular hodographs. Therefore it follows that an inverse square force generates elliptical orbits.

More generally it can be proved that Newton's laws, plus a central force proportional to the inverse square of the distance, generate orbits that are conic sections. It can be shown that the position of the point S, i.e. the common origin of the velocity vectors, with respect to the circle centered in C discriminates the orbit. In fig. 3.31 (a) we see the case discussed of the point S within the circle. When the point S is exactly on the circumference, then the orbit is a parabola, while if the point is external to the circle, the orbit is an hyperbola. Fig. 3.31 panel (c) shows that there is an angle for which the tangent $\overline{ST}$ is perpendicular to the radius $\overline{CT}$. There is another specular point T' on the left for which the same condition applies. When the orbiting planet is in this condition it means that the change in velocity is in the same direction as the velocity or, in other words, the satellite is moving on a straight line. On this specific orbit the planet is coming from infinity along the direction parallel to $\overline{ST}$, swings around the Sun, and proceeds towards infinity approaching more and more the line parallel to $\overline{ST'}$. The parabola is the limiting case in which the planet will reach infinity with zero velocity. In fact, the parabola hodograph has maximum perihelion v_p and zero aphelion v_a velocities.

3.2.4 Kepler's Problem: Newton's Solution Using Calculus

In this section we work out the solutions to Kepler problems using the *standard* treatment given in the majority of undergraduate physics books, i.e. we will use calculus. We cannot certainly exhaustively cover the theory and practice of differential calculus: instead, we will give hints on the various techniques, always trying to be self-consistent.

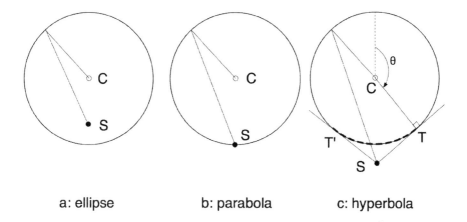

a: ellipse   b: parabola   c: hyperbola

FIGURE 3.31 The hodographs for ellipse (a), parabola (b), and hyperbola (c).

Let us show how to prove that the ellipse is the curve that a planet follows in its orbit around the Sun. We assume that the gravitational force is central and is proportional to the inverse square of the distance between the Sun and the planet. We assume we have enough experimental evidence to state that the force between the Sun and a planet is given by eq. 2.61. In the following equations we indicate vectors with bold characters like, for example, velocity **v**. Scalars will be indicated with italics to signify that they are just numbers. Finally, unit vectors will be indicated by a hat symbol over the quantity. For example, a velocity vector aligned along the x axis can be written as $\mathbf{v} = v\,\hat{\mathbf{x}}$ where v is the magnitude of the velocity. The time derivative will be indicated either with the symbol $\frac{d}{dt}$ or with a dot above the quantity to be derived. A double dot means second derivative. Sometimes we might find it convenient to mix the two notations.

We have already stated that the gravitational attraction is a central force and therefore it depends only on the distance between the Sun and the planet. If **r** is the radius vector joining the Sun with the planet, the central force F can be written as:

$$\mathbf{F} = F(r)\hat{\mathbf{r}} \tag{3.96}$$

where $\hat{\mathbf{r}}$ is the unit vector aligned as **r**. A planet of mass $m \ll M$, where M is the mass of the Sun, subjected to the gravitational force will obey Newton's second law:

$$m\ddot{\mathbf{r}} = F(r)\hat{\mathbf{r}} \tag{3.97}$$

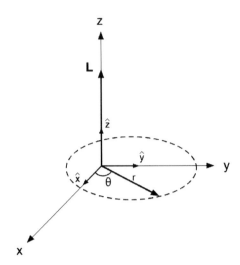

FIGURE 3.32 Angular momentum vector **L** perpendicular to the plane of the orbit. Note the unit vectors $\hat{x}$, $\hat{y}$, and $\hat{z}$.

This is the equation that we need to solve to obtain the orbit of the planet around the Sun.

We have already encountered a quantity that we called *angular momentum* (eq. 3.24), which is conserved in a gravitational orbit. Let us study the concept more in detail. We call **angular momentum** with respect to a point O in space the following quantity:

$$\mathbf{L} = \mathbf{r} \times \mathbf{p} = \mathbf{r} \times m\mathbf{v} = \mathbf{r} \times m\dot{\mathbf{r}} \tag{3.98}$$

where $\mathbf{p} = m\mathbf{v} = m\dot{\mathbf{r}}$ is the linear momentum of the planet of mass m and O is the reference point. Since the gravitational attraction is central, if we choose the point O to coincide with the position of the Sun, then (1) the angular momentum is conserved and (2) the orbit is planar, i.e. the planet orbits in a plane passing through O and perpendicular to the vector **L**. To show this, let's take the vector product of eq. 3.97:

$$\mathbf{r} \times m\ddot{\mathbf{r}} = \mathbf{r} \times F(r)\hat{\mathbf{r}} = F(r)\ \mathbf{r} \times \hat{\mathbf{r}} = 0 \tag{3.99}$$

The vector product $\mathbf{r} \times \hat{\mathbf{r}} = 0$ because the two vectors are parallel and therefore $\sin \alpha$ in eq. 2.35 is equal to zero. We can write:

$$\frac{d\mathbf{L}}{dt} = \frac{d}{dt}(\mathbf{r} \times m\dot{\mathbf{r}}) = m(\mathbf{r} \times \ddot{\mathbf{r}}) + m(\dot{\mathbf{r}} \times \dot{\mathbf{r}}) = 0$$
$$\mathbf{L} = \mathbf{r} \times m\dot{\mathbf{r}} = \text{constant} \tag{3.100}$$

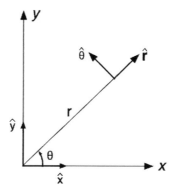

FIGURE 3.33 Polar coordinate system showing the unit vectors $\hat{r}$ and $\hat{\theta}$.

Eq. 3.100 tells us that the derivative of the angular momentum is zero and therefore is a conserved quantity and this proves point (1) above.

To prove point (2) we now take the scalar product with $\mathbf{r}$ of the angular momentum:

$$\mathbf{r} \cdot \mathbf{L} = 0 \tag{3.101}$$

because $\mathbf{L}$ is the vector product of $\mathbf{r}$, $\dot{\mathbf{r}}$ is perpendicular to the plane identified by these two vectors and so is perpendicular to both. So $\mathbf{L}$ and $\mathbf{r}$ are always perpendicular: the orbit is confined to the plane containing $\mathbf{r}$ and $\dot{\mathbf{r}}$ through O. This proves point (2) above (see fig. 3.32).

We have therefore established that the orbit under a central force is confined to be on a plane through O. It is now natural to choose a coordinate system with origin at O where the Sun is located and where we use polar coordinates (r, θ) with $\mathbf{r}$ spanning the plane perpendicular to $\mathbf{L}$.

Let's equip ourselves with some useful vector relationships. In fig. 3.33 we see that we defined the two unit vectors $\hat{r}$ and $\hat{\theta}$ in the direction of increasing r and θ. We have already discussed how to rotate vectors by a certain angle. Eq. 2.8 gives us the recipe to rotate a vector when expressed as a 2-component column vector. Since $\hat{r}$ and $\hat{\theta}$ are perpendicular, from fig. 3.33 we see that we can obtain the coordinates of the rotated vector by using eq. 2.8:

$$\begin{pmatrix} \hat{r} \\ \hat{\theta} \end{pmatrix} = \begin{pmatrix} \cos\theta & \sin\theta \\ -\sin\theta & \cos\theta \end{pmatrix} \begin{pmatrix} \hat{x} \\ \hat{y} \end{pmatrix} \tag{3.102}$$

Remember that eq. 3.102 is a compact way to write the components of the rotated vectors as:

$$\hat{r} = \begin{pmatrix} \cos\theta \\ \sin\theta \end{pmatrix}$$

$$\hat{\theta} = \begin{pmatrix} -\sin\theta \\ \cos\theta \end{pmatrix}$$

(3.103)

which simply means that the new unit vectors can be expressed as:

$$\hat{r} = \cos\theta \; \hat{x} + \sin\theta \; \hat{y}$$

$$\hat{\theta} = -\sin\theta \; \hat{x} + \cos\theta \; \hat{y}$$

(3.104)

Using matrix notation we can write the vector **r** as:

$$\mathbf{r} = r \begin{pmatrix} \cos\theta \\ \sin\theta \end{pmatrix} = r \; \hat{\mathbf{r}}$$

(3.105)

Studying the motion of the planet means studying the time derivatives of the various vectors involved. Let's start with the vector **r** and let's express the derivative using the matrix notation in eq. 3.105. We need to establish how to take derivatives of column vectors. The rule is simple: the derivative of a vector is a new vector whose components are the derivative of the vector components being derived. Given a generic vector **w**:

$$\mathbf{w} = \begin{pmatrix} w_1 \\ w_2 \end{pmatrix}$$

(3.106)

Its derivative is:

$$\dot{\mathbf{w}} = \begin{pmatrix} \dot{w}_1 \\ \dot{w}_2 \end{pmatrix} = \begin{pmatrix} \frac{dw_1}{dt} \\ \frac{dw_2}{dt} \end{pmatrix}$$

(3.107)

If w_1 and w_2 are themselves functions $w_1 = w_1(x)$ and $w_2 = w_2(x)$, then we can apply the chain rule and we have:

$$\dot{\mathbf{w}}(x) = \begin{pmatrix} \frac{dw_1}{dx}\frac{dx}{dt} \\ \frac{dw_2}{dx}\frac{dx}{dt} \end{pmatrix} = \dot{x} \begin{pmatrix} \frac{dw_1}{dx} \\ \frac{dw_2}{dx} \end{pmatrix}$$

(3.108)

Now that we have the recipe to derive column vectors, let's take the time derivative of the vector 3.105:

$$\frac{d\mathbf{r}}{dt} = \frac{d}{dt}\left[r \begin{pmatrix} \cos\theta \\ \sin\theta \end{pmatrix}\right] = \dot{r} \begin{pmatrix} \cos\theta \\ \sin\theta \end{pmatrix} + r\dot{\theta} \begin{pmatrix} -\sin\theta \\ \cos\theta \end{pmatrix}$$

(3.109)

Using eq. 3.103, we have:

$$\dot{\mathbf{r}} = \dot{r}\hat{\mathbf{r}} + +r\dot{\theta}\hat{\boldsymbol{\theta}}$$

(3.110)

Eq. 3.110 gives us the polar components of the velocity in polar coordinates. The time variation of the vector $\mathbf{r}$ can be decomposed in a **radial component** and a **transverse component** (see fig. 3.33). The radial component is $\dot{r}$ and the transverse component is $r\dot{\theta}$.

The components of the acceleration can also be expressed in terms of the polar unit vectors. Taking the derivative of eq. 3.109 we have:

$$\ddot{\mathbf{r}} = \ddot{r}\begin{pmatrix} \cos\theta \\ \sin\theta \end{pmatrix} + \dot{r}\dot{\theta}\begin{pmatrix} -\sin\theta \\ \cos\theta \end{pmatrix} + \dot{r}\dot{\theta}\begin{pmatrix} -\sin\theta \\ \cos\theta \end{pmatrix} + \dot{r}\ddot{\theta}\begin{pmatrix} -\sin\theta \\ \cos\theta \end{pmatrix} + r\dot{\theta}^2\begin{pmatrix} -\cos\theta \\ -\sin\theta \end{pmatrix} \tag{3.111}$$

Collecting the terms:

$$\ddot{\mathbf{r}} = (\ddot{r} - r\dot{\theta}^2)\begin{pmatrix} \cos\theta \\ \sin\theta \end{pmatrix} + (2\dot{r}\dot{\theta} + r\ddot{\theta})\begin{pmatrix} -\sin\theta \\ \cos\theta \end{pmatrix} \tag{3.112}$$

or equivalently:

$$\ddot{\mathbf{r}} = (\ddot{r} - r\dot{\theta}^2)\hat{\mathbf{r}} + (2\dot{r}\dot{\theta} + r\ddot{\theta})\hat{\boldsymbol{\theta}} \tag{3.113}$$

Let's look at the second term on the right-hand side of eq. 3.113. We notice that it can be written as:

$$2\dot{r}\dot{\theta} + r\ddot{\theta} = \frac{1}{r}\frac{d}{dt}(r^2\dot{\theta}) \tag{3.114}$$

Using eq. 3.114, eq. 3.113 can be written as:

$$\ddot{\mathbf{r}} = (\ddot{r} - r\dot{\theta}^2)\hat{\mathbf{r}} + \frac{1}{r}\frac{d}{dt}(r^2\dot{\theta})\hat{\boldsymbol{\theta}} \tag{3.115}$$

Eqs. 3.110 and 3.115 give us the radial and traverse components of velocity and acceleration. Let's now insert the expression of $\ddot{\mathbf{r}}$ we just found into the equation of motion of the planet 3.97.

$$m\ddot{\mathbf{r}} = F(r)\hat{\mathbf{r}}$$
$$m[(\ddot{r} - r\dot{\theta}^2)\hat{\mathbf{r}} + \frac{1}{r}\frac{d}{dt}(r^2\dot{\theta})\hat{\boldsymbol{\theta}}] = F(r)\hat{\mathbf{r}} \tag{3.116}$$

The above vector equation, i.e. an equation containing vector quantities, can be re-expressed in terms of relations between the scalar components of the vectors. In fact, two vectors are equal *if and only if* their components are equal. In the case of the second equation 3.116, the scalar components of $\hat{\mathbf{r}}$ on the left-hand side must equal the scalar components of $\hat{\mathbf{r}}$ on the right-hand side. Same for $\hat{\boldsymbol{\theta}}$. The vector eq. 3.116 becomes two scalar equations:

$$m(\ddot{r} - r\dot{\theta}^2) = F(r)$$
$$\frac{m}{r}\frac{d}{dt}(r^2\dot{\theta}) = 0 \tag{3.117}$$

These two equations can be solved by finding the two functions of time $r = r(t)$ and $\theta = \theta(t)$ which satisfy eq. 3.117. The orbit can then be reproduced as follows[17]: for each time t, we can plot the point $P = P(r, \theta)$. Varying the time t moves the point P which describes the orbit.

With a little bit of algebra we can reduce the two eqs. 3.117 to a single equation whose solution is directly the equation of the orbit $r = r(\theta)$. The second equation in 3.117 immediately gives us the conservation of angular momentum, which we write as:

$$\dot{\theta} = \frac{h}{r^2} \tag{3.118}$$

where the scalar h is the angular momentum per unit mass $h = \frac{L}{m}$ (see eq. 3.24). We can now eliminate $\dot{\theta}$ from the first equation in 3.117 which becomes:

$$m\left(\ddot{r} - \frac{h^2}{r^3}\right) = F(r) \tag{3.119}$$

Let's now insert Newton's gravitational force expression for $F(r)$:

$$m\left(\ddot{r} - \frac{h^2}{r^3}\right) = -G\frac{Mm}{r^2} \tag{3.120}$$

This differential equation is quite difficult to solve in this form. We will show that it is possible to transform eq. 3.120 into a simpler equation in some other function u such that we can easily apply the techniques to solve ordinary second-order differential equations.

To this end, let us define a new function $u = \frac{1}{r}$. We will show that such a choice simplifies the equations. In fact, taking into account that if $r = r(\theta)$ then we must have that $u = u(\theta)$ and the derivatives are:

$$\dot{r} = -\frac{1}{u^2}\frac{du}{dt} = -\frac{1}{u^2}\frac{du}{d\theta}\dot{\theta} = -\frac{1}{u^2}\frac{du}{d\theta}hu^2 = -h\frac{du}{d\theta}$$

$$\ddot{r} = \frac{d}{dt}\left(-h\frac{du}{d\theta}\right) = -h\dot{\theta}\frac{d^2u}{d\theta^2} = -h^2u^2\frac{d^2u}{d\theta^2} \tag{3.121}$$

We now insert the expression for $\ddot{r}$ into eq. 3.120:

$$m\left(\ddot{r} - \frac{h^2}{r^3}\right) = -G\frac{Mm}{r^2}$$

$$m\left(-h^2u^2\frac{d^2u}{d\theta^2} - \frac{h^2}{r^3}\right) = -GMmu^2 \tag{3.122}$$

$$mh^2u^2\frac{d^2u}{d\theta^2} + mh^2u = -GMmu^2$$

Dividing by mh^2 we finally have the equation:

---

[17]This method of representing curves is called *parametric representation*.

$$\frac{d^2u}{d\theta^2} + u = \frac{GM}{h^2} \qquad (3.123)$$

This is a second-order constant coefficients inhomogeneous differential equation.

The solution involves two steps. Step (1): find the solution to the homogeneous equation, i.e. the equation without the constant term:

$$\frac{d^2u}{d\theta^2} + u = 0 \qquad (3.124)$$

The solution to eq. 3.124 is called the *complementary function* u_{CF}. Step (2): find a *particular solution* u_{PS} which, in the case of our equation 3.123, is simply $u_{PS} = \frac{GM}{h^2}$ as can be seen by direct substitution. The general solution is then $u = u_{CF} + u_{PS}$.

In order to find the complementary function, we make use of an important property of the solutions of linear homogeneous second-order differential equations: if $u_1(\theta)$ and $u_2(\theta)$ are linearly independent solutions[18], then the general solution is:

$$u_{CF} = C_1 u_1 + C_2 u_2 \qquad (3.125)$$

where C_1 and C_2 are two arbitrary constants that must be given or determined by the physics of the problem.

If we look at the homogeneous equation 3.124 we see immediately that the trigonometric functions sine and cosine are solutions because deriving twice they return the same function with a changed sign. So the general solution to the homogeneous equation is:

$$u_{CF} = C_1 \cos\theta + C_2 \sin\theta \qquad (3.126)$$

Using the trigonometric identity:

$$\cos(\alpha - \beta) = \cos\alpha\cos\beta + sin\alpha sin\beta \qquad (3.127)$$

and taking $C_1 = A\cos\theta_0$ and $C_2 = A\cos\theta_0$, we have that eq. 3.126 can be written as:

$$u_{CF} = A\cos(\theta - \theta_0) \qquad (3.128)$$

and the general solution is:

$$u = A\cos(\theta - \theta_0) + \frac{GM}{h^2} \qquad (3.129)$$

θ_0 is the initial angle that for simplicity we can assume to be zero so the

---

[18]The functions $u_1(\theta)$ and $u_2(\theta)$ are linearly independent if one function is not a multiple of the other function.

planet starts its revolution around the Sun at $\theta = 0$. We can finally bring back the radius vector $r = \frac{1}{u}$ and we have:

$$
\begin{aligned}
\frac{1}{r} &= A\cos(\theta - \theta_0) + \frac{GM}{h^2} \\
\frac{h^2}{GM}\frac{1}{r} &= A\cos(\theta - \theta_0) + 1
\end{aligned}
\tag{3.130}
$$

or, extracting r we finally have the equation of the orbit:

$$
r = \frac{\frac{h^2}{GM}}{1 + e\cos\theta}
\tag{3.131}
$$

where we indicated the constant A with the symbol e.

Eq. 3.130 is the polar form of a conic section as proved by eq. 3.77. Comparing the two equations we see immediately that the semi-latus rectum is:

$$
p = a(1 - e^2) = \frac{h^2}{GM}
\tag{3.132}
$$

We have finally proved that a conic section is the orbit of a planet of mass m subject to a central force proportional to the inverse square of the distance to the Sun.

3.2.5 Kepler's Problem: Solution Using Geometric Algebra with the Laplace-Runge-Lenz Vector

We have seen two different ways to prove the inverse Kepler's problem. We now introduce a third one. We left for the finale what we think is the most elegant solution. We will make use of what is known as **geometric algebra** (GA), i.e. a different way to use geometric objects like, points, lines, planes, etc. The new approach consists of using the geometric objects as members of an algebra[19] without relying on equations between the coordinates of the objects. It follows that GA *does not use coordinates*. Once we have defined what the objects of the GA are, geometric operations on these objects like, for example rotations, are done by using the algebraic operations instead of equations involving the coordinates. However, we will find that sometimes we need to use coordinates in some situations.

Let us try to justify why we can use GA instead of using the more familiar vector algebra. When we introduced operations with vectors in section 2.1.2 (page 56), the vector product (or cross product) was defined as an operation between two vectors $\mathbf{a}$ and $\mathbf{b}$[20] producing another vector $\mathbf{c}$ perpendicular to

---

[19]Let us briefly define an algebra: a set of objects where two operations are defined (addition and multiplication). The mathematical reader will have noticed that we have just defined a *ring* in mathematics.

[20]We continue to use the notation whereas vectors are indicated with boldface symbols.

both **a** and **b**. Because of this definition, **c** cannot belong to the plane defined by **a** and **b**. So, the vector **c** extends the plane into a 3rd dimension. We have also seen that the vector **c** is not exactly like the two vectors **a** and **b**, but is what we called a *pseudovector* because of the anti-symmetric property $\mathbf{a} \times \mathbf{b} = -\mathbf{b} \times \mathbf{a}$. We say that the 2D vector space is not *closed* with respect to cross product, i.e. the result of the vector product produces an object that does not belong to the original space where the vectors **a** and **b** "live". It would be desirable to have a definition of multiplication between vectors that does not require adding an extra dimension. In order to do so, we need to define how we multiply the unit vectors **x** and **y**. Given a vector **x**, the metric axiom of GA is:

$$\text{Metric axiom:} \quad \mathbf{xx} = \mathbf{x}^2 = \|x\|^2 \tag{3.133}$$

From the metric axiom 3.133, it follows that:

$$\hat{\mathbf{x}}\hat{\mathbf{x}} = \hat{\mathbf{x}}^2 = 1 \tag{3.134}$$

$\hat{\mathbf{x}}$ (and $\hat{\mathbf{y}}$) that are the unit vectors perpendicular to each other.

Another important consequence of the metric axiom derives from the Pythagorean Theorem applied to the vector $\hat{\mathbf{x}} + \hat{\mathbf{y}}$ sum of the two unit vectors, which has magnitude $\sqrt{2}$. Since its magnitude is $\|\hat{\mathbf{x}} + \hat{\mathbf{y}}\|^2 = 2$ we must have:

$$\begin{aligned} \|(\hat{\mathbf{x}} + \hat{\mathbf{y}})\|^2 &= \|(\hat{\mathbf{x}} + \hat{\mathbf{y}})(\hat{\mathbf{x}} + \hat{\mathbf{y}})\|^2 \\ &= \|\hat{\mathbf{x}}\hat{\mathbf{x}} + \hat{\mathbf{x}}\hat{\mathbf{y}} + \hat{\mathbf{y}}\hat{\mathbf{x}} + \hat{\mathbf{y}}\hat{\mathbf{y}}\|^2 \\ &= \|1 + \hat{\mathbf{x}}\hat{\mathbf{y}} + \hat{\mathbf{y}}\hat{\mathbf{x}} + 1\|^2 \\ &= 2 \end{aligned} \tag{3.135}$$

Equation 3.135 is true if and only if:

$$\mathbf{xy} = -\mathbf{yx} \tag{3.136}$$

Let us try to see what happens if we multiply two vectors algebraically. Suppose we have two vectors $\mathbf{a} = a_x\hat{\mathbf{x}} + a_y\hat{\mathbf{y}}$ and $\mathbf{b} = b_x\hat{\mathbf{x}} + a_b\hat{\mathbf{y}}$. And let's multiply them using the usual algebra:

$$\begin{aligned} \mathbf{xy} &= (a_x\hat{\mathbf{x}} + a_y\hat{\mathbf{y}})(b_x\hat{\mathbf{x}} + b_y\hat{\mathbf{y}}) \\ &= a_xb_x\hat{\mathbf{x}}\hat{\mathbf{x}} + a_xb_y\hat{\mathbf{x}}\hat{\mathbf{y}} + a_yb_x\hat{\mathbf{y}}\hat{\mathbf{x}} + a_yb_y\hat{\mathbf{y}}\hat{\mathbf{y}} \\ &= (a_xb_x + a_yb_y) + (a_xb_y - a_yb_x)\hat{\mathbf{x}}\hat{\mathbf{y}} \end{aligned} \tag{3.137}$$

If we multiply the coordinates algebraically, we end up with an interesting relation:

$$\mathbf{xy} = (a_xb_x + a_yb_y) + (a_xb_y - a_yb_x)\hat{\mathbf{x}}\hat{\mathbf{y}} \tag{3.138}$$

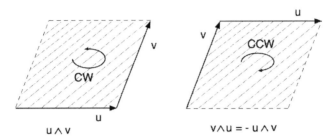

FIGURE 3.34 Geometric interpretation of the wedge product as an oriented area. The left panel shows that the orientation $\mathbf{u} \wedge \mathbf{v}$ is opposite to $\mathbf{v} \wedge \mathbf{u} = -\mathbf{u} \wedge \mathbf{v}$.

The first term on the right-hand side is the scalar (dot) product of the two vectors, while the second term reminds us of the $z-$component of the vector (cross) product, i.e. the area of the parallelogram built with the two vectors $\mathbf{a}$ and $\mathbf{b}$. The first object is a scalar quantity (number) while the second quantity is a new object that we call **bivector** and we indicate the new operation with the *wedge* symbol $\wedge$. Therefore we have a new definition of the product of two vectors:

$$\mathbf{xy} = \mathbf{x} \cdot \mathbf{y} + \mathbf{x} \wedge \mathbf{y} \qquad (3.139)$$

The geometric product of two vectors in eq. 3.139 produces a scalar and a bivector. So if we define the algebra to contain these objects, then the algebra is closed, i.e. the outputs of the multiplication are elements within the original space. The wedge product has an immediate geometric interpretation which is partially shared with the conventional vector product: the wedge product represents the area of the parallelogram built with the two vectors. The sign depends on the orientation of the area as shown in fig. 3.34.

The basic blocks of this new 2D algebra therefore are: scalars, called **grade 0**, 2 orthogonal unit vectors $\hat{\mathbf{x}}$ and $\hat{\mathbf{y}}$, called **grade 1**, and bivectors, called **grade 2**. This classification is analogous to the more familiar classification of standard geometrical objects where a point has zero dimension, a line is 1-dimensional and a plane is 2-dimensional. Any object in this GA algebra can be expressed as a combination of these basic blocks. These new objects, combinations of the basic blocks, are called **multivectors** and we indicate them with capital letters. An example of a multivector can be written as:

$$A = 1 + 3\hat{\mathbf{x}} - 5\hat{\mathbf{y}} - 7\hat{\mathbf{x}} \wedge \hat{\mathbf{y}} \qquad (3.140)$$

where we see that all the basic blocks have been used. Multiplication of multivectors generates other multivectors so the algebra is closed.

An interesting object is the square of the product of two unit vectors:

$$(\hat{\mathbf{x}}\hat{\mathbf{y}})^2 = (\hat{\mathbf{x}}\hat{\mathbf{y}})(\hat{\mathbf{x}}\hat{\mathbf{y}})$$
$$= -\hat{\mathbf{y}}\hat{\mathbf{x}}\hat{\mathbf{x}}\hat{\mathbf{y}}$$
$$= -\hat{\mathbf{y}}\hat{\mathbf{x}}^2\hat{\mathbf{y}} \tag{3.141}$$
$$= -\hat{\mathbf{y}}\hat{\mathbf{y}}$$
$$= -1$$

Eq. 3.141 tells us that the square of the bivector $\hat{\mathbf{x}}\hat{\mathbf{y}}$ is equal to -1, indicating a possibly close connection with the imaginary number $i = \sqrt{-1}$. In other words, pure geometry shows that the imaginary numbers have a geometric meaning. In close analogy we define:

$$\hat{\mathbf{x}}\hat{\mathbf{y}} = I \tag{3.142}$$

This bivector is connected with rotations just like imaginary number i. Let's see what happens if we multiply a vector $\mathbf{u} = a\hat{\mathbf{x}} + b\hat{\mathbf{y}}$ by I **from the right**:

$$\mathbf{u}' = \mathbf{u}I = (a\hat{\mathbf{x}} + b\hat{\mathbf{y}})\hat{\mathbf{x}}\hat{\mathbf{y}}$$
$$= a\hat{\mathbf{x}}\hat{\mathbf{x}}\hat{\mathbf{y}} + b\hat{\mathbf{y}}\hat{\mathbf{x}}\hat{\mathbf{y}} \tag{3.143}$$
$$= -b\hat{\mathbf{x}} + a\hat{\mathbf{y}}$$

It can be easily shown that the new vector $\mathbf{u}' = \mathbf{u}I$ is obtained by rotating $\mathbf{u}$ by 90° CCW. The vector $\mathbf{u}' = I\mathbf{u}$ instead rotates by 90° CW.

We now turn our attention again to the motion of a planet in the solar system. We now show that GA has many benefits over the more traditional approach of previous sections.

Let's define a bivector:

$$H = \mathbf{r} \wedge \dot{\mathbf{r}} \tag{3.144}$$

This vector is the angular momentum per unit mass L/m. Let's verify that it is conserved in the presence of a central force.

$$\frac{dH}{dt} = \frac{d}{dt}(\mathbf{r} \wedge \dot{\mathbf{r}}) = \dot{\mathbf{r}} \wedge \dot{\mathbf{r}} + \mathbf{r} \wedge \ddot{\mathbf{r}} = 0 \tag{3.145}$$

H is conserved because the wedge product of a vector by itself is zero and $\mathbf{r}$ is parallel to $\ddot{\mathbf{r}}$ because of the central force being in the same direction as $\mathbf{r}$. We have seen that this conservation law implies that the orbital motion is confined to a plane and that the radius vector $\mathbf{r}$ sweeps equal areas in equal times.

The definition of unit vector a is $\hat{\mathbf{r}} = \frac{\mathbf{r}}{r}$, so we can express the vector $\mathbf{r} = r\hat{\mathbf{r}}$. Eq. 3.144 becomes:

$$H = (r\hat{\mathbf{r}}) \wedge (\dot{r}\hat{\mathbf{r}} + r\dot{\hat{\mathbf{r}}})$$
$$= \cancel{(r\hat{\mathbf{r}}) \wedge (\dot{r}\hat{\mathbf{r}})} + (r\hat{\mathbf{r}}) \wedge (r\dot{\hat{\mathbf{r}}})$$
$$= r^2(\hat{\mathbf{r}} \wedge \dot{\hat{\mathbf{r}}}) \qquad (3.146)$$
$$= r^2(\hat{\mathbf{r}}\dot{\hat{\mathbf{r}}} - \hat{\mathbf{r}} \cdot \dot{\hat{\mathbf{r}}})$$
$$= r^2\hat{\mathbf{r}}\dot{\hat{\mathbf{r}}}$$

where we have used the definition of a geometric algebra product of two vectors and the fact that $\hat{\mathbf{r}} \cdot \dot{\hat{\mathbf{r}}} = 0$.[21]

We now want to show how to obtain the shape of the orbit of a planet orbiting the Sun. The planet is subject to the force:

$$\mathbf{f} = \ddot{\mathbf{r}} = -\frac{GM}{r^2}\hat{\mathbf{r}} \qquad (3.147)$$

Let's multiply the two equations 3.146 and 3.147:

$$H\ddot{\mathbf{r}} = r^2\hat{\mathbf{r}}\dot{\hat{\mathbf{r}}}(-\frac{GM}{r^2})\hat{\mathbf{r}}$$
$$= -GM\hat{\mathbf{r}}\dot{\hat{\mathbf{r}}}\hat{\mathbf{r}} \qquad (3.148)$$

The two vectors $\hat{\mathbf{r}}$ and $\dot{\hat{\mathbf{r}}}$ are perpendicular and therefore anti-commute. Eq. 3.148 becomes:

$$H\ddot{\mathbf{r}} = GM\dot{\hat{\mathbf{r}}}\hat{\mathbf{r}}\hat{\mathbf{r}}$$
$$= GM\dot{\hat{\mathbf{r}}}\hat{\mathbf{r}}^2 = GM\dot{\hat{\mathbf{r}}} \qquad (3.149)$$

Let's rewrite eq. 3.149 as:

$$H\ddot{\mathbf{r}} - GM\dot{\hat{\mathbf{r}}} = 0 \qquad (3.150)$$

Since we proved that H is constant, we notice that eq. 3.150 can be written as:

$$\frac{d}{dt}(H\dot{\mathbf{r}} - GM\hat{\mathbf{r}}) = 0$$
$$A = H\dot{\mathbf{r}} - GM\hat{\mathbf{r}} = \text{constant} \qquad (3.151)$$

The vector A is a constant of motion, i.e. it stays the same no matter where it is calculated along the orbit. This vector is called the **Laplace-Runge-Lenz** (LRL) vector. From the LRL vector it is straightforward to obtain the equation of the orbit and therefore solve the inverse Kepler problem.

---

[21]$\hat{\mathbf{r}} \cdot \dot{\hat{\mathbf{r}}} = \frac{1}{2}\frac{d}{dt}(\hat{\mathbf{r}} \cdot \hat{\mathbf{r}}) = \frac{1}{2}\frac{d}{dt}(\|\hat{\mathbf{r}}\|^2) = 0.$

Let's rewrite eq. 3.151 as:

$$H\dot{\mathbf{r}} = A + GM\hat{\mathbf{r}} \tag{3.152}$$

and multiply by $\mathbf{r}$:

$$H\dot{\mathbf{r}}\mathbf{r} = A\mathbf{r} + GM\hat{\mathbf{r}}\mathbf{r} \tag{3.153}$$

and expand the products:

$$H(\dot{\mathbf{r}} \cdot \mathbf{r} + \dot{\mathbf{r}} \wedge \mathbf{r}) = A \cdot \mathbf{r} + A \wedge \mathbf{r} + GM\hat{\mathbf{r}} \cdot \mathbf{r} + \cancel{GM\hat{\mathbf{r}} \wedge \mathbf{r}} \tag{3.154}$$

We know that $H = \mathbf{r} \wedge \dot{\mathbf{r}}$ and so $\dot{\mathbf{r}} \wedge \mathbf{r} = -H$. We also know that $\hat{\mathbf{r}}$ and $\mathbf{r}$ are parallel and therefore their scalar product is simply r. We therefore simplify eq. 3.154 into:

$$(\dot{\mathbf{r}} \cdot \mathbf{r})H - H^2 = Ar \cos\theta + A \wedge \mathbf{r} + GMr \tag{3.155}$$

Let us discuss the various terms in eq. 3.155. The first term in the left-hand side is a bivector, the second term in the left-hand side is a scalar. The three terms on the right hand side are respectively a scalar, a bivector, and a scalar. Eq. 3.155 equates two multivectors and they are equal if and only if separately the various components are equal, i.e. the scalars on the right hand side must be equal to the scalars on the left-hand side, and the same for the vectors and bivectors.

Equating the scalars in eq. 3.155 we have:

$$H^2 = Ar \cos\theta + GMr \tag{3.156}$$

where the angle θ is the angle between the LRL vector A and the radius vector r. Eq. 3.156 can be written in a more familiar way:

$$r = \frac{\frac{H^2}{GM}}{1 + \frac{A}{GM} \cos\theta} \tag{3.157}$$

The orbit described in eq. 3.157 is identical to the orbit obtained in the previous section in eq. 3.131. We see that the eccentricity is $e = \frac{A}{GM}$ while the scalar H^2 is exactly equal to h^2.

3.3   ENERGY AND ORBITS

In section 3.1.1 we have seen that if a dynamical system is invariant under rotations and time translations, then it conserves, respectively, angular momentum and energy. A planet of mass m, subject to a central force proportional to the inverse square of the distance, orbiting the Sun of mass M (where $m << M$), falls under this category. The total conserved energy of the planet can be written as:

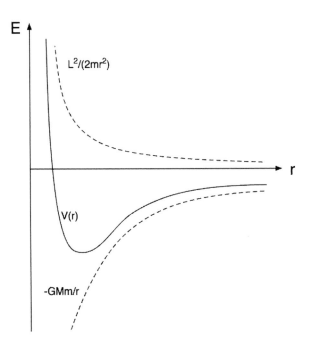

FIGURE 3.35 Potential energy of a planet orbiting the Sun. The two components, centrifugal and gravitational, are plotted as dashed lines.

$$E = \frac{1}{2}m(\dot{r}^2 + r^2\dot{\theta}^2) - \frac{GMm}{r} \tag{3.158}$$

Using the expression of the angular momentum given by eq. 3.85, eq. 3.158 can be written as:

$$E = \frac{1}{2}m\dot{r}^2 + \frac{1}{2}\frac{L^2}{mr^2} - \frac{GMm}{r} \tag{3.159}$$

where we use the fact that the potential associated with the gravitational force is:

$$V(r) = \frac{d}{dr}\frac{-GMm}{r^2} = -\frac{GMm}{r} \tag{3.160}$$

The total energy is the sum of three components: kinetic energy, a term containing the total angular momentum, and the potential energy. We already know that the third term is the gravitational potential whose derivative is the conservative gravitational force. In complete analogy, we can express the second term as a potential U_c:

$$U_c = \frac{1}{2}\frac{L^2}{mr^2} \tag{3.161}$$

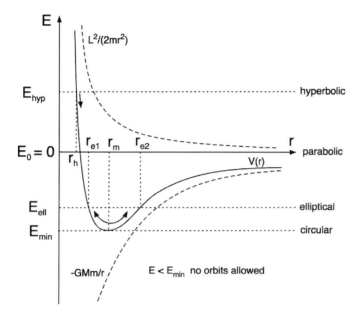

FIGURE 3.36 Relationship between orbits and energy.

whose derivative produces an associated force equal to:

$$F_c = -\frac{\partial U_c}{\partial r} = -\frac{\partial}{\partial r}\left(\frac{1}{2}\frac{L^2}{mr^2}\right) = \frac{L^2}{mr^3} = mr\dot{\theta}^2 \qquad (3.162)$$

This force is called **centrifugal force** and is directed *from* the Sun *to* the planet exactly opposite to the gravitational attraction. The potential 3.161 can be regarded as a centrifugal potential energy. If we include it with the gravitational potential energy $U(r)$ we have an *effective potential energy* for central forces:

$$V(r) = -\frac{GMm}{r} + \frac{L^2}{2mr^2} \qquad (3.163)$$

Notice that the centrifugal potential *reduces* the effect of the gravitational potential because of the opposite signs in the forces. The potential $V(r)$ is plotted in fig. 3.35 as a continuous line, while the two components of the effective potential are plotted as dashed lines. There are a few interesting facts to underline: first, the potential rises up at small r proportional to r^{-2}, meaning that the centrifugal force becomes stronger than the attractive gravitational force acting effectively as a repulsive force. Second, the potential has a minimum, therefore excluding orbits below the corresponding energy for a given angular momentum. Third, for large r the gravitational potential dominates over the centrifugal term.

We can see very clearly the relationship between the energy of a planet and its possible orbits (see fig. 3.36). By studying the plot of the effective potential we can see the relationship between the energy and the type of conic section. In fig. 3.163 we plot again the effective potential, but now we study the consequences of changing the total energy of the planet. For total energy $E_{hyp} > 0$, a planet coming from the right ($r = \infty$) in the plot has enough energy to approach the Sun up to a minimum distance $r = r_h$ and then fly back to reach infinity. The special case $E = 0$ represents a planet coming from infinity with exactly zero energy and flying back to infinity along a parabolic trajectory. As the energy becomes negative, the planet is bound and confined between the two distances $r_{e1} < r_{e2}$ with oscillatory motion. The corresponding orbit is an ellipse. The final special case is a planet that has the minimum energy allowed for which the planet's orbit at a fixed minimum distance $r = r_m$ corresponds to a circular orbit.

3.4 THE UNIVERSAL LAW OF GRAVITATION: ONE VERY FAMOUS APPLE

In the previous sections we have described in some detail the relationship between Newton's laws of motion, Kepler's laws, and Newton's law of gravitation. We have seen that Newton's laws are somehow more basic than Kepler's laws in that these last can be obtained by the former. Newton's law of gravitation has been introduced without a good discussion because we were interested mostly in the mathematical structure. We now go into the foundation of the fantastic intuition that Newton had when he made gravity a **universal** force of attraction between masses.

In 1665 and 1666 the bubonic plague, also-called black death, spread in London forcing a lot of people to leave the city and repair in the isolation of small villages hoping that the disease would not reach them. The University of Cambridge was also closed and young Isaac Newton was sent 94 miles north of London to his native village of Woolsthorpe-by-Colsterworth. Like one of the Greek philosophers, Newton had all the time to ponder about maths and physics. The legend says that one day Newton noticed an apple falling to the ground and this put in motion a series of speculations. First, according to his second law of motion $F = ma$, since the apple is accelerating when falling, it means that there is a force attracting the apple to the ground and this force is proportional to the mass of the apple. Newton was interested also in another problem. Why is the Moon going around the Earth? According to his first law, the Moon should have left the Earth unless a force is keeping it circling around in a circular orbit. He knew it was more or less a circular orbit because the apparent diameter of the Moon does not change with time. We can imagine how Newton might have reasoned at this point: there is a force that is attracting the apple to the ground. Could it be the same force that is attracting the Moon to the Earth? Additionally, why doesn't the Moon fall to the Earth and destroy it?

In order to prove this hypothesis, Newton made some calculations. He knew that the orbit of the Moon is almost circular since its apparent diameter does not change with time and he knew that the period of revolution of the Moon around the Earth is about 27.3 days (lunar month). He also knew that the distance from the Earth to the Moon is about 31 times the diameter of the Earth (see Aristarchus's estimate in chapter 1). We can immediately calculate the speed of the Moon while orbiting the Earth:

$$v = \frac{\pi D_{EM}}{4T} = 3680 \text{ km/h} \tag{3.164}$$

where $D_{EM} = 384,400$ km is the distance from the Earth to the Moon and $T = 27.3$ days is the period of revolution of the Moon around the Earth. We know that a body on a circular motion at constant velocity is subject to an acceleration $\sim v^2/R$ which, in the case of the Moon is:

$$a_M = \frac{v^2}{D_{EM}} = 0.0027 \text{ m/s}^2 \tag{3.165}$$

Newton also knew about the acceleration of a free-falling body at the surface of the Earth. This is usually indicated with g and is equal to 9.81 m/s². This value is much larger than the value calculated for the Moon and this fact sparked an idea in Newton[22]. Is it conceivable that the acceleration of the Moon is due to the attraction of the Earth and is much smaller because the Moon is far away? What kind of relationship with distance must such force have in order to explain the values of g and a_M?

The suggestion that the Earth's attraction is responsible for both the falling of an apple and the fact that the Moon stays in orbit around the Earth is quite bold. Can it be proved experimentally, perhaps checking that effectively g is different between different heights? In 1662 and 1665 Hooke reported to the Royal Society the results of a series of experiments conducted on the towers of Westminster Abbey and old St. Paul's Cathedral. Hooke did not find evidence of difference in the force of gravity between weights whose heights were 90 feet apart. Negative results notwithstanding, Hooke effectively proposed an inverse square law and Newton must have known about Hooke's hypothesis.

Let's return to Newton's calculations. If the Moon is 62 times more distant than the apple, then the two accelerations must be in the ratio:

---

[22]Newton was not the first to come up with the idea of a universal force of gravitation. Robert Hooke (1635-1703) was a contemporary of Newton and claimed that he was the original proposer of the inverse square law for the gravitational attraction. A long and bitter dispute between the two scientists started when Newton published his work on the spectral content of white light in 1672. Hooke dismissed Newton's work on optics and this fact certainly upset Newton who, probably as a retaliation, did not mention Hooke's work in his *Principia*.

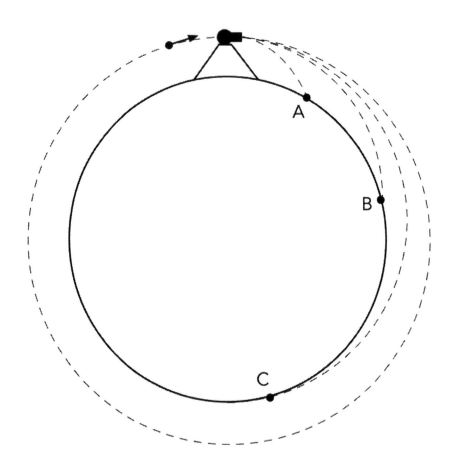

FIGURE 3.37 Newton's cannonball thought experiment where cannon-balls are shot horizontally with different velocities. For velocities less than a special value, the cannonball will fall on Earth, although at increasing distances A,B,C with increasing the velocity. There is a special velocity for which the cannonball does not fall on Earth but keeps falling constantly and eventually will hit the cannon from behind!

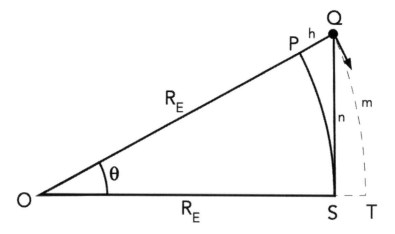

FIGURE 3.38 Geometry to estimate the orbital speed in Newton's cannonball thought experiment.

$$\frac{a}{g} = \frac{1}{62^2} \rightarrow a = \frac{g}{62^2} = 0.0026 \text{ m/s}^2 \tag{3.166}$$

This value is in good agreement with the value estimated in eq. 3.165. So it seems that the force attracting the apple on the surface of the Earth is the same force that attracts the Moon to the Earth. There are two things to still discuss: why does the Moon not fall to the Earth like the apple? Second, and most important, the estimates above assume that all the mass of the Earth is located at its center as depicted in fig. 3.39.

The first question has been brilliantly answered by Newton. He seems to have been the first to realize that the action of a force *perpendicular* to the direction of motion will change the direction but not the magnitude of the velocity vector. So, qualitatively, the Moon does not fall on the Earth because its velocity vector is perpendicular to the force of attraction of the Earth. Newton then performs a **thought experiment**, i.e. a mental image of an experiment difficult or impossible to realize in practice, where a hypothesis is tested and the consequences are analyzed. He imagined to be on the top of a mountain where a cannon horizontally shoots cannonballs with different velocities as depicted in fig. 3.37. Increasing the horizontal velocity of the cannonball has the effect of increasing the distance where the cannonball lands. Landing points A, B and C correspond to increasing horizontal velocities. Newton then brings the argument further and realizes that if the horizontal speed is high enough, the cannonball does not land but keeps circling the Earth. In other words, the cannonball keeps falling but it keeps the same altitude over the Earth's surface: the curvature of the cannonball trajectory matches *exactly* the curvature of the surface of the Earth. This fact allows us to calculate the orbital speed if we know the radius of the Earth R_E, the height of the mountain h, and gravity acceleration g.

Fig. 3.38 shows the geometry of a cannonball shot horizontally with respect to the Earth's surface (the arc PS). The cannon is located at height $h = \overline{PQ}$ and shoots horizontally in the direction of the vector indicated. The condition for a closed circular orbit requires that the cannonball has exactly the necessary velocity such that the curved path $m = \overline{QT}$ is exactly parallel to the curved path $\overline{PS}$ representing the surface of the Earth. Suppose that the mountain $\overline{PQ}$ has an altitude of $h = 8,000$ meters and that the radius of the Earth $R_E = 6,371$ km. The acceleration due to gravity at Q is $h = 9.7783$ m/s^2 which is about 0.3% less than the acceleration at sea level. If the cannonball has no horizontal speed, the free-fall time from Q to P is given by:

$$h = \frac{1}{2}gt^2$$
$$t = \sqrt{\frac{2h}{g}} = 40.4 \text{ sec}$$

(3.167)

Since $h \ll R_E$ we can assume that the segment $n = \overline{QS} \approx m = \overline{QT}$. We have:

$$m \approx n = \sqrt{(R_E + h)^2 - R_E^2} = \sqrt{h^2 + 2hR_E} = 319.4 \text{ km}$$
$$v_{orb} = \frac{319.5}{40.4} = 7,906 \text{ m/s}$$

(3.168)

The orbital velocity calculated in eq. 3.168 gives the correct value as can be verified by using the usual formula $v_{orb} = \sqrt{\frac{GM}{R_E}}$.

3.4.1 Newton's Shell Theorem Using Calculus

The second question, also known as *Newton's shell theorem*, is quite important and requires a bit of effort. The universal law of gravitation between two planets is expressed by eq. 2.61 that we conveniently report here:

$$\vec{F}_{12} = G\frac{m_1 m_2}{r_{12}^2}\vec{k}_{12}$$

(3.169)

Eq. 3.169 contains the distance r_{12}^2 representing the distance between the centers of the two planets of masses m_1 and m_2. Eq. 3.169 assumes that the two planets have *spherical symmetry* and have densities depending only on r. Before publishing his *Principia*, Newton had to prove the above statement,

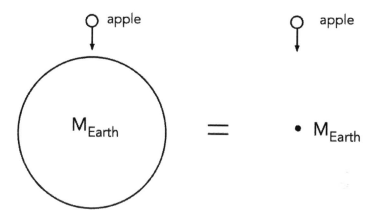

FIGURE 3.39 The force of gravitational attraction of the Earth on an apple. On the left, the force is due to the distributed mass of the Earth. On the right, the mass of the Earth is all concentrated in its center. Are the two forces equal?

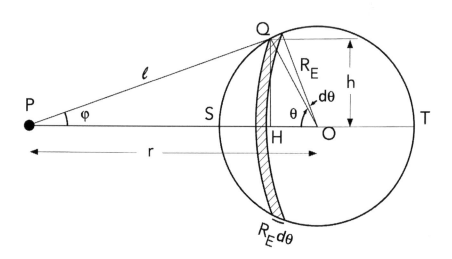

FIGURE 3.40 Geometry for proving Newton's shell theorem using calculus.

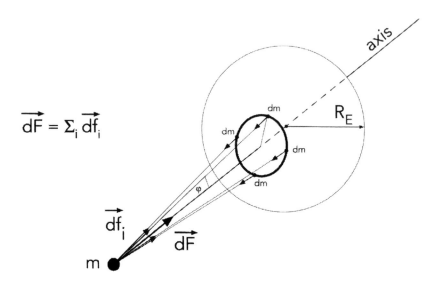

$$\vec{dF} = \Sigma_i \, \vec{df_i}$$

FIGURE 3.41 Force dF exerted by a ring of infinitesimal thickness along the symmetry axis passing through the center of the sphere and the point mass m. Each of the df_i is pointing towards an infinitesimal mass dm around the ring and they are all disposed on a cone of semi-angle φ. The projection of each df_i on the symmetry axis is $df_i \cos \varphi$.

which he did using a geometric argument. We will show the proof of the shell theorem by using various methods.

Let us start with the *modern* proof based on calculus (see fig. 3.40). Consider a spherical shell of infinitesimal thickness of total mass M and consider a point mass m located at a distance r from the center of the spherical shell. We want to calculate the gravitational force exerted on the mass m at point P by the thin shell. In order to do this we divide the shell in an infinite number of rings of infinitesimal size (shaded area in fig. 3.40). Once we calculate the force exerted by the ring, we sum over all the rings obtained by moving the point Q from S to T. When we do this, we see that the angle θ varies from 0 to 2π, the angle φ varies from 0 to a max value corresponding to $\theta = \pi/2$, and ℓ increases from a minimum value equal to $r - R_E$ to a max value equal to $r + R_E$. From simple geometry, we know that the total area of the sphere of radius R_E is $4\pi R_E^2$. We immediately see that the *mass density* σ of the shell is the total mass M divided by the total surface area $4\pi R_E^2$:

$$\sigma = \frac{M}{4\pi R_E^2} \qquad (3.170)$$

The force dF exerted by the infinitesimal ring is oriented along the symmetry axis and is obtained by projecting each df_i, i.e. by multiplying by $\cos \varphi$ (see fig. 3.41). Using Newton's universal gravitational formula, we have:

$$dF = \frac{Gm \, dM}{\ell^2} \cos \varphi \qquad (3.171)$$

where dM is the mass of the infinitesimal ring and the $\cos \varphi$ factor takes into account the symmetrical disposition of all the df_i. We need to calculate the mass dM of the infinitesimal ring and then add all the rings to cover all the surface of the planet. Going back to fig. 3.40, we see that the shaded area of the ring is equal to $R_E \theta$, so the infinitesimal area of the ring will be:

$$\begin{aligned} dA &= 2\pi h \cdot R_E d\theta \\ &= 2\pi R_E^2 \cdot \sin \theta d\theta \end{aligned} \qquad (3.172)$$

The mass of the ring is simply the product of the mass surface density 3.170 times the area dA 3.172:

$$\begin{aligned} dM &= \sigma \cdot dA \\ &= \sigma \cdot 2\pi R_E^2 \cdot \sin\theta d\theta \\ &= \frac{M}{4\pi R_E^2} \cdot 2\pi R_E^2 \cdot \sin\theta d\theta \\ &= \frac{1}{2} M \sin \theta d\theta \end{aligned} \qquad (3.173)$$

Eq. 3.171 becomes:

$$dF = \frac{Gm}{\ell^2} \cos \varphi \cdot \frac{1}{2} M \sin \theta d\theta \qquad (3.174)$$

We now need to integrate all the dF corresponding to all the rings covering the sphere in fig. 3.40.

$$F = \int \frac{Gm}{\ell^2} \cos \varphi \cdot \frac{1}{2} M \sin \theta d\theta \qquad (3.175)$$

Both $\cos \varphi$ and ℓ depend on θ. In order to evaluate the integral 3.175 we now show that it is possible to eliminate θ and φ, so that the integrand in 3.175 depends only on ℓ. Let's look again at fig. 3.40. We can apply the law of cosines to the triangle $\triangle PQO$ in two different ways:

$$R_E^2 = r^2 + \ell^2 - 2r\ell \cos \varphi \qquad (3.176)$$

from which we obtain:

$$\cos \varphi = \frac{r^2 + \ell^2 - R_E^2}{2r\ell} \qquad (3.177)$$

The other law of cosines is:

$$\ell^2 = r^2 + R_E^2 - 2rR_E \cos\theta \tag{3.178}$$

We now differentiate both terms of eq. 3.178:

$$2\ell d\ell = -2rR_E(-\sin\theta d\theta) \tag{3.179}$$

from which we extract:

$$d\theta = \frac{\ell d\ell}{rR_E \sin\theta} \tag{3.180}$$

Inserting eq. 3.180 and eq. 3.177 into eq. 3.175 and simplifying all the common terms, we have:

$$
\begin{aligned}
F &= \int_{r-R_E}^{r+R_E} \frac{Gm}{\ell^2} \cos\varphi \cdot \frac{1}{2} M \sin\theta d\theta \\
&= \frac{GmM}{4r^2 R_E} \int_{r-R_E}^{r+R_E} \frac{r^2 + \ell^2 - R_E^2}{\ell^2} d\ell \\
&= \frac{GmM}{4r^2 R_E} \int_{r-R_E}^{r+R_E} \left(1 + \frac{r^2 - R_E^2}{\ell^2}\right) d\ell
\end{aligned}
\tag{3.181}
$$

where the integral in eq. 3.181 now depends only on the variable ℓ, which we have already shown varies between $r - R_E$ and $r + R_E$. We now use the Fundamental Theorem of Calculus , expressed in eq. 1.31, and look for the function whose derivative is equal to the integrand, i.e. the function in parenthesis in eq. 3.181.

It is easy to verify that the function:

$$G(\ell) = \ell - \frac{r^2 - R_E^2}{\ell} \tag{3.182}$$

satisfies the condition. Therefore, using eq. 1.37 we have:

$$
\begin{aligned}
F &= \frac{GmM}{4r^2 R_E} \int_{r-R_E}^{r+R_E} \frac{r^2 + \ell^2 - R_E^2}{\ell^2} d\ell \\
&= \frac{GmM}{4r^2 R_E} (G(r + R_E) - G(r - R_E)) \\
&= \frac{GmM}{4r^2 R_E} \left(r + R_E - \frac{r^2 - R_E^2}{r + R_E} - \left(r - R_E - \frac{r^2 - R_E^2}{r - R_E}\right)\right) \\
&= \frac{GmM}{4r^2 R_E} \cdot 4R = \frac{GmM}{r^2}
\end{aligned}
\tag{3.183}
$$

which proves that the force on a mass m at a distance r from a spherical

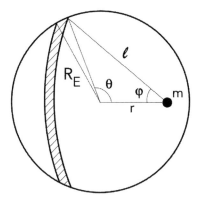

FIGURE 3.42 Geometry to prove Newton's shell theorem when the mass m is inside the shell.

shell can be calculated by considering the mass of the shell collapsed into its center. We can now imagine that a solid spherical body can be decomposed into an infinite number of concentric shells. In this case, each shell of total mass dM can be collapsed to the common center and the sum of all shells will correspond to the total mass of the planet being collapsed into its center. It is also evident that the argument is valid if and only if the density of the planet is a function of r, its distance from the center, or in other words, each infinitesimal shell has a constant surface density.

It is now simple to calculate the force on the point mass m if it is inside the shell. In this case, fig. 3.40 is modified into fig. 3.42. If the point mass m is inside the shell, we now see that the variable ℓ is varying between $R_E + r$ and $R_E - r$.

Eq. 3.183 needs to be evaluated inserting the new integration limits:

$$
\begin{aligned}
F &= \frac{GmM}{4r^2 R_E} \int_{R_E - r}^{R_E + r} \frac{r^2 + \ell^2 - R_E^2}{\ell^2} d\ell \\
&= \frac{GmM}{4r^2 R_E} (G(R_E + r) - G(R_E - r)) \\
&= \frac{GmM}{4r^2 R_E} \left(R_E + r - \frac{r^2 - R_E^2}{R_E + r} - \left(R_E - r - \frac{r^2 - R_E^2}{R_E - r} \right) \right) \\
&= \frac{GmM}{4r^2 R_E} \cdot (R_E + r - (R_E - r) - (r - R_E - (-(r - R_E)))) = 0
\end{aligned}
$$

(3.184)

The total gravitational force on a mass m inside a shell is zero.

3.4.2   Newton's Shell Theorem Using Geometry

After having gone through the *standard* proof of Newton's shell theorem using calculus, let's now discuss how Newton himself proved it using a geometrical construction[23]. As is well known, Newton did not use calculus in his proofs, even though he just invented it! Instead, he relied on geometrical arguments coupled to his laws of motion and gravitation. Newton proved his shell theorem in the year 1685 and he inserted his proof in the *Principia* as Proposition 71 of Book 1.

Without diminishing at all Newton's greatness, we have to say that the proof as reported in the *Principia* is not straightforward. Therefore we will spend some extra effort in order to make sure that the reader appreciates in full the beauty of geometry applied to physics. We now proceed to calculate the gravitational force exerted by an *extended* spherical body on a point mass m. Newton assumed that the spherical body was made up of an infinite collection of infinitesimal particles whose density was only a function of the radius R_E of the body. He then proceeded to calculate the force exerted by each particle and summed vectorially over all the particles. Let's consider fig. 3.43, which uses the same exact notation used by Newton. In the top panel we have a point mass m located at the point P belonging to the plane passing through the center of the shell S. The bottom panel is exactly the same but this time the point mass m is located closer to the center S. We consider a ring obtained by rotating the arc HI around the axis PS and the corresponding ring obtained by rotating the arc hi around the axis pS. We want to show that the ratio of the gravitational force due to the HI ring (F_1) to the gravitational force due to the hi ring (F_2) is equal only to $\overline{pS}^2/\overline{PS}^2$, i.e. depends only on the inverse of the distance squared to the center of the shell.

It is absolutely critical that the reader understands the way in which Newton built the geometry shown in fig. 3.43, taken straight from Newton's *Principia*, including the letters labelling the various points. As we have already discussed in the previous section, we want to calculate the gravitational force at Point P due to the matter belonging to a ring obtained by rotating the small arc IH around the axis PB. We then calculate the force at point p due to the matter belonging to a ring obtained by rotating the small arc ih around the axis pb. In the limit in which the chord IH is small and can be approximated with the segment $\overline{IH}$, we have that the area of the ring obtained rotating around the axis PB is:

$$A_P = 2\pi \cdot \overline{IH} \cdot \overline{IQ} \tag{3.185}$$

And, if we consider the bottom panel in fig. 3.43, we have that the corresponding area is:

$$A_p = 2\pi \cdot \overline{ih} \cdot \overline{iq} \tag{3.186}$$

---

[23]We will present an extended discussion following the work of Weinstock [11].

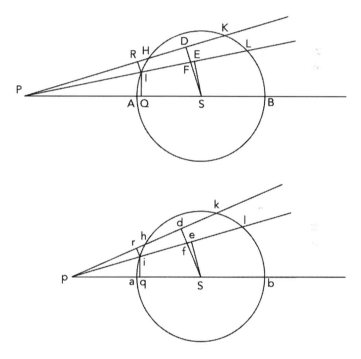

FIGURE 3.43 Newton's geometry of his proof of the shell theorem when the mass m is outside the shell.

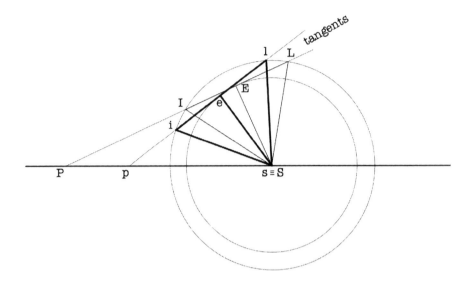

FIGURE 3.44 Auxiliary geometry to show that the segments in fig. 3.43 obey to the equalities $SE = se$ and $IL = il$.

Due to the symmetry of the ring with respect to the axis where the points P and p are placed, the corresponding forces that the rings exert on a point mass m of unit mass is[24]:

$$F_1 = GM \frac{A_P}{\overline{PI}^2} \frac{\overline{PQ}}{\overline{PI}}$$
$$F_2 = GM \frac{A_p}{\overline{pi}^2} \frac{\overline{pq}}{\overline{pi}}$$

(3.187)

where we used Newton's universal law of gravitation. The factors $\frac{\overline{PQ}}{\overline{PI}}$ and $\frac{\overline{pq}}{\overline{pi}}$ take into account the cosine of the angles $\angle KPB$ and $\angle kpb$ in fig. 3.43.

We want to show that the ratio $F_1/F_2 \propto \overline{ps}^2/\overline{PS}^2$. To this end we need to carefully study fig. 3.43. Newton built the figure 3.43 very cleverly: he traced the line $PIEL$ and he divided the chord IL into two segments IE and EL such that the segment SE is exactly perpendicular to $PIEL$. Now the clever bit: he chose another point p along the same axis and such that the line $piel$ identifies the chord il and the segment se, perpendicular to $piel$. It turns out that you can build the segment iel equal in length to the segment IEL and the segment se equal to the segment SE and both perpendicular to respectively $PIEL$ and $piel$. In fig. 3.44 we show in fact that the triangle

---

[24]The force per unit mass is the field intensity.

$\triangle ISL$ is obtained by rotating the bold triangle $\triangle isl$. The two triangles are in fact congruent because the segment $\overline{SE}$ is equal to the segment $\overline{se}$ being two radii of the same smaller circle. The segment $\overline{SL}$ is equal to the segment $\overline{sl}$ because they are the radii of the same larger circle and, by construction, the angles $\angle SEL$ and $\angle sel$ are both right angles. It follows that the two triangles $\triangle ISL$ and $\triangle isl$ are *congruent* as has been discussed earlier in this chapter (SAS condition in fig. 3.17). The exact same reasoning can be applied to the triangles $\triangle SHD$ and $\triangle shd$. Therefore we have the first set of equalities:

$$\overline{se} = \overline{SE}$$
$$\overline{sd} = \overline{SD} \tag{3.188}$$

If the segments $\overline{IH}$ and $\overline{ih}$ are small, then we can assume that:

$$\overline{PF} \simeq \overline{PE}$$
$$\overline{SF} \simeq \overline{SE}$$
$$\overline{pf} \simeq \overline{pe}$$
$$\overline{sf} \simeq \overline{se} \tag{3.189}$$

Looking again at fig. 3.43 we can write now a chain of equalities:

$$\overline{df} = \overline{sd} - \overline{sf} = \overline{sd} - \overline{se} = \overline{SD} - \overline{SE} = \overline{SD} - \overline{SF} = \overline{DF} \tag{3.190}$$

Newton now uses similarity between triangles. Let us refer to the fig. 3.45 where we isolated the two triangles $\triangle PES$ and $\triangle PIQ$. By construction, these are both right triangles because the point I is projected onto the point Q and the point S is projected onto the point E. They also share the same angle $\angle EPS$ and, as a result, the last angles $\angle PIQ$ and $\angle PSE$ are equal.

Therefore the two triangles are similar having all angles equal. It follows that we can write the following relations:

$$\frac{\overline{PI}}{\overline{PS}} = \frac{\overline{IQ}}{\overline{SE}} \tag{3.191}$$

and

$$\frac{\overline{PI}}{\overline{PS}} = \frac{\overline{PQ}}{\overline{PE}} = \frac{\overline{PQ}}{\overline{PF}} \tag{3.192}$$

Similarly, for the lower case triangles, we have:

$$\frac{\overline{pi}}{\overline{ps}} = \frac{\overline{iq}}{\overline{se}} \tag{3.193}$$

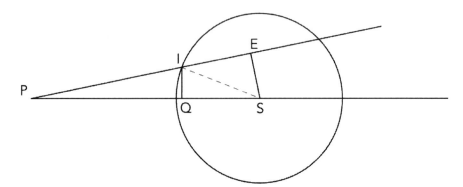

FIGURE 3.45 Auxiliary geometry to exploit similarities between triangles $\triangle PES$ and $\triangle PIQ$.

and

$$\frac{\overline{pi}}{\overline{ps}} = \frac{\overline{pq}}{\overline{pe}} = \frac{\overline{pq}}{\overline{pf}} \qquad (3.194)$$

We now turn our attention to two similar triangles $\triangle PIR$ and $\triangle PFD$ as shown in fig. 3.46.

We have:

$$\frac{\overline{PI}}{\overline{PF}} = \frac{\overline{RI}}{\overline{DF}} \qquad (3.195)$$

and, for the lowercase:

$$\frac{\overline{pf}}{\overline{di}} = \frac{\overline{df}}{\overline{ri}} \qquad (3.196)$$

We can now multiply the left-hand side of the first equation with the left-hand side of the second equation in eq. 3.195:

$$\frac{\overline{PI}}{\overline{PF}} \cdot \frac{\overline{pf}}{\overline{pi}} = \frac{\overline{RI}}{\overline{DF}} \cdot \frac{\overline{df}}{\overline{ri}} = \frac{\overline{RI}}{\overline{ri}} \qquad (3.197)$$

We need two more figures to study before proceeding to the final calculation. Let us now consider fig. 3.47 showing the two triangles $\triangle IRH$ and $\triangle IES$. Notice that we expanded the triangle $\triangle IRH$ to show the difference between the arc IH and the segment $\overline{IH}$. In the limit of the point H approaching point I, the line $\overline{HI}$ approaches better and better, the tangent at point I.

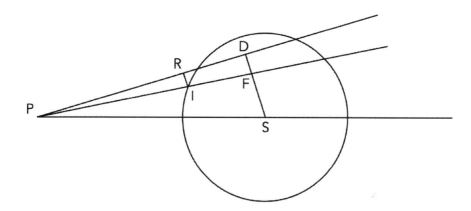

FIGURE 3.46 Auxiliary geometry to exploit similarities between triangles $\triangle PIR$ and $\triangle PFD$.

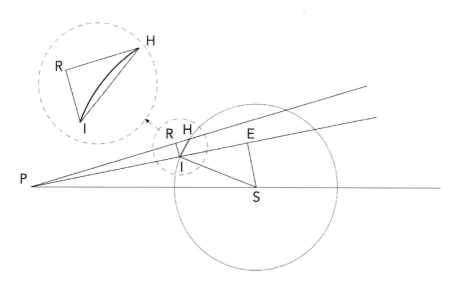

FIGURE 3.47 Auxiliary geometry to show the relationships between triangles $\triangle IRH$ and $\triangle IES$.

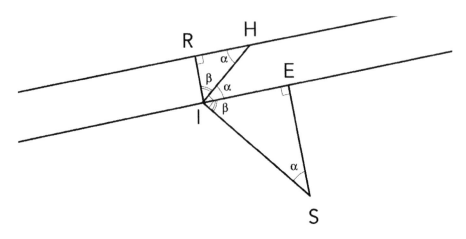

FIGURE 3.48 Auxiliary geometry to show that the triangles $\triangle IRH$ and $\triangle IES$ are similar.

This means that, for angle $\angle RPI$ sufficiently small, the triangle $\triangle IRH$ can be approximated as a right triangle where the segment $\overline{HI}$ is the hypothenuse perpendicular to the radius $\overline{SI}$. Now we already know that, by construction, $\overline{RI}$ is perpendicular to $\overline{PH}$ and therefore, in the small arc approximation, $\overline{RI}$ is perpendicular to $\overline{IE}$.

Now the key point: when the point H approaches the point I, the two lines PH and PE become more and more parallel. In the limit, we can consider them parallel as shown in fig. 3.48. In this figure, the line HI cuts the two parallel lines RH and IE. This means that the angle $\alpha = \angle RHI$ is equal to the angle $\angle ISE$. Since the angle $\angle HIS$ is a right angle as is $\angle RIE$, it follows that the angle $\beta = \angle RIH$ is equal to the angle $\angle EIS$. The two triangles $\triangle IRH$ and $\triangle IES$ are similar because they have all their angles equal. If we look again at fig. 3.44 we see that the triangle $\triangle IES$ is congruent to triangle $\triangle ies$, therefore triangle $\triangle irh$ is similar to triangle $\triangle IRH$. We can finally write:

$$\frac{\overline{RI}}{\overline{ri}} = \frac{\overline{HI}}{\overline{hi}} \tag{3.198}$$

and eq. 3.197 becomes:

$$\frac{\overline{pf}}{\overline{PF}} \frac{\overline{PI}}{\overline{pi}} = \frac{\overline{HI}}{\overline{hi}} \tag{3.199}$$

We have all the ingredients to evaluate the ratio of the two forces F_1/F_2. We will show that $F_1/F_2 = \overline{ps}^2/\overline{PS}^2$. Using eq. 3.187 we have:

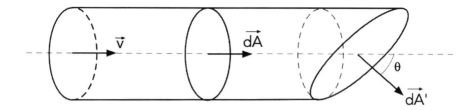

FIGURE 3.49 Fluid flowing at constant velocity through an infinitesimal pipe of cross section dA.

$$\frac{F_1}{F_2} = \frac{\overline{\frac{PQ \cdot HI \cdot IQ}{\overline{PI}^3}}}{\overline{\frac{\overline{pq} \cdot hi \cdot iq}{\overline{pi}^3}}} = \frac{\overline{PQ} \cdot \overline{HI} \cdot \overline{IQ}}{\overline{pq} \cdot \overline{hi} \cdot \overline{iq}} \cdot \frac{\overline{pi}^3}{\overline{PI}^3} \tag{3.200}$$

Inserting in eq. 3.200 $\overline{IQ}$ from eq. 3.191, $\overline{iq}$ from eq. 3.193, $\overline{PQ}$ from eq. 3.192, and $\overline{pq}$ from eq. 3.194, we have:

$$\frac{F_1}{F_2} = \frac{\overline{PI}}{\overline{PS}}\frac{\overline{PF}}{\overline{pq}} \cdot \frac{\overline{HI}}{\overline{hi}} \cdot \frac{\overline{PI}}{\overline{PS}}\frac{\overline{SE}}{\overline{iq}} \cdot \frac{\overline{pi}^3}{\overline{PI}^3}$$

$$= \frac{\overline{PF}}{\overline{pf}}\frac{\overline{pi}}{\overline{PI}} \cdot \frac{\overline{HI}}{\overline{hi}} \cdot \frac{\overline{SE}}{\overline{se}} \cdot \frac{\overline{ps}^2}{\overline{PS}^2} \tag{3.201}$$

Finally, using eq. 3.188 and eq. 3.199, we have:

$$\frac{F_1}{F_2} = \frac{\overline{ps}^2}{\overline{PS}^2} \tag{3.202}$$

which shows that the ratio of the forces at P and p are inversely proportional to the distance to the **center of the shell**. Newton then claims that, given the symmetry, the same result is obtained if we consider the whole shell and all the concentric shells until we consider the whole planet. In other words, the force on a particle external to a spherical body can be calculated assuming the total mass is concentrated in the center of the body.

3.4.3 Newton's Shell Theorem Using Gauss's Law

In this section we will derive Newton's shell theorem using an elegant theorem due to Gauss and Ostrogradsky called the *divergence theorem*. Gauss's theorem is beautiful, elegant, and useful in many diverse fields, from electromagnetism to fluid dynamics. As is customary in this book, we will introduce all the concepts needed for full comprehension of the theorem first.

Gauss's law expresses a mathematical relationship between the flux of a

FIGURE 3.50 Carl Friedrich Gauss was a German mathematician and physicist who made substantial contributions to algebra, astronomy, and electromagnetism.

vector field out of a surface enclosing a tridimensional volume and the source of the field, i.e. the distribution of masses inside the tridimensional volume[25].

We need to introduce the concept of *field* and the associated *flux* out of a surface. A field (in physics) is a region of space where a physical quantity is associated with each of the points of the region. For example, the temperature of a room can be represented by a *temperature field* $T = T(x, y, z)$ where a temperature T is associated with each point (x, y, z) in Cartesian coordinates. In this case we say that the temperature T is a *scalar* field because the temperature is a scalar quantity, i.e. is a number. Another important scalar field is the density, i.e. the number of particles per unit volume $\rho = \rho(x, y, z)$ at each of the points in space (x, y, z).

In complete analogy we can define a *vector field* $\overrightarrow{F} = \overrightarrow{F}(x, y, z)$ where a vector is associated with each point in space. An important example is the gravitational field[26], i.e. the force that a mass $M >> m$ exerts on a unit mass

---

[25]Gauss's theorem is heavily used in electromagnetism, where instead of *masses* we have electric *charges*.

[26]Here the treatment is purely Newtonian, which is an approximation of a more accurate theory of gravitation due to A. Einstein and called the *General Theory of Relativity*. The gravitational field is described by the field equations of general relativity that equate the Einstein tensor G to the stress-energy tensor T. While in Newtonian gravity the gravitational field depends only on the spatial distribution of matter, in general relativity the field depends on the distribution of matter and energy.

m. Depending on where in space we put our unit test mass m, this will feel a force (per unit mass):

$$\vec{g}(r) = -\frac{\vec{F_g}}{m} = -\frac{GM}{r^2}\hat{r} \tag{3.203}$$

where $r = \sqrt{x^2 + y^2 + z^2}$. $\vec{g}(r)$ is the *gravitational field* at the point r and $\hat{r}$ is the unit vector directed *from* the test particle *to* the mass generating the gravitational field.

The air velocity in a room is another example of a vector field because at each point in space is associated a vector indicating direction and speed of the air. The fields just described are *stationary*, i.e. the values they have at the specified points in space do not depend on time. More in general, the fields might depend on time so that a generic field can be written as $f = f(x, y, z, t)$.

Having defined a vector field we now turn our attention to the concept of *flux* of a vector field through a surface. Fig. 3.49 shows a pipe with infinitesimal circular cross section. Inside the pipe we have a fluid flowing from left to right at constant velocity $\vec{v}$ which we define as positive. Imagine that, on the other end of the pipe, there is a bucket collecting the fluid. We can associate the concept of flux, in this example, with the amount of fluid collected per unit time. What is the amount of water per unit time? It can be found by multiplying the fluid velocity times the area $d\vec{A}$ as in fig. 3.49. How do we choose the area $d\vec{A}$? It seems obvious that the "correct" area to choose is the one parallel to the velocity $\vec{v}$. If we want to make the calculation more general, we can choose any area $d\vec{A'}$ oriented at angle θ with respect to the velocity vector. However, in order to make sure that the flux calculated, i.e. the amount of water collected stays the same, we need to adjust for the fact that $d\vec{A'}$ is bigger than $d\vec{A}$ and tilted by an angle θ.

It is easy to verify that the (infinitesimal) flux $d\Phi$ of the velocity vector can be written as a dot product:

$$d\Phi = \vec{v} \cdot d\vec{A'} \tag{3.204}$$

We now need to define properly what we mean with the symbol $d\vec{A'}$. What does it mean, a vector associated with an area? We can associate a vector to an area by defining:

$$d\vec{A} = \vec{n}\, dA \tag{3.205}$$

where $\vec{n}$ is a unit vector perpendicular to the area dA. There is still an ambiguity to resolve: any surface has two faces and therefore two perpendicular vectors. When we decide the orientation of the unit vector $\vec{n}$, we break the ambiguity and we talk about *oriented* area. In the case of the flux above, we decided that the area dA is oriented as the vector $\vec{v}$ so that the flux is defined as a positive quantity. The other choice will obviously give a negative flux.

The common convention is to define the unit vector $\vec{n}$ *outwards* from any *closed* surface.

With this definition of the area as a vector, we can rewrite eq. 3.204 as:

$$d\Phi = (\vec{v} \cdot \vec{n})dA' \tag{3.206}$$

where now dA' is a scalar quantity equal to the infinitesimal area.

If the surface for which we want to calculate the flux is finite (see fig. 3.51) we can proceed by dividing the surface in many small surfaces dA_i with associated unit vectors n_i. Obviously now the vector $\vec{v}$ must be evaluated at each area element and therefore it will be now a set of vectors $\vec{v_i}$ associated with the respective infinitesimal area elements dA_i. The finite flux Φ is the sum over all the small areas:

$$\Phi = \sum_i (\vec{v_i} \cdot \vec{n_i})dA_i' \tag{3.207}$$

What we just described is a special kind of integral called a *surface integral* and is written as:

$$\Phi = \oint_S (\vec{v} \cdot \vec{n})dA \tag{3.208}$$

where the symbol $\oint_S$ in eq. 3.207 represents the limit for the areas dA tending to zero over the whole surface S.

Let's turn to physics and suppose we have a point source of incompressible fluid (for example, water) inside a big container. Let's suppose the source is outputting F cubic meters of water per second. We assume that the water exits the point source equally in all directions, i.e. *isotropically* (see fig. 3.52). If the closed spherical surface S completely surrounds the source, then the water exiting the surface must equal the water exiting the point source[27]. Here we see the importance of the fluid being incompressible. The requirement of isotropic output flow can be restated by assuming that the velocity of the water depends only on the radius $v = v(\vec{r})$. We can now use eq. 3.208:

$$F = \oint_S (\vec{v} \cdot \vec{n})dA = \oint_S \vec{v} \cdot d\vec{A} = v \oint_S dA \tag{3.209}$$

The last equality in eq. 3.209 follows from the fact that the velocity vector of the water $\vec{v}$ is perpendicular to the area element $d\vec{A}$. In the case of isotropic flow, the velocity v is constant at the surface of the sphere and therefore can be put outside the integral sign.

Eq. 3.209 tells us something very interesting: the value of F depends only on the presence of the source inside the closed surface S. It is important that

---

[27]If the source is not isotropic, the amount of fluid exiting the closed surface is still equal to the amount of fluid generated by the source. In this case the velocity vector is not constant on the surface of the sphere.

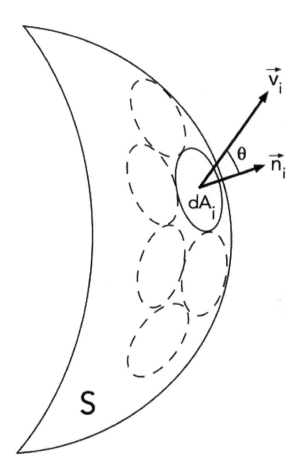

FIGURE 3.51 Representation of the flux of a vector $\vec{v}$ through a surface S.

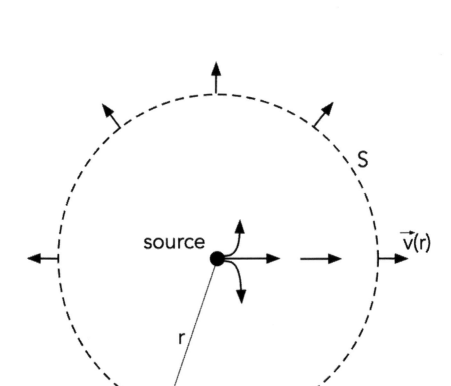

FIGURE 3.52 Water source inside a spherical closed surface S.

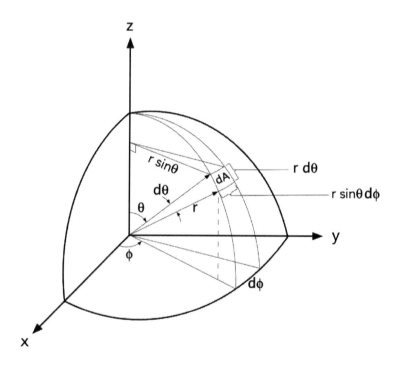

FIGURE 3.53 Geometry to determine the area element dA of S.

the vector $\vec{v} = \vec{v}(r)$ is a function only of the radius r, so that it has a constant value on the surface of the sphere S and points radially out.

We are left with estimating the value of the integral $\oint_S dA$. The little square dA in fig. 3.53 is obtained by multiplying the height $rd\theta$ by the base $r\sin\theta d\theta$. The whole surface of the sphere is described when the angle ϕ is allowed to range between 0 and 2π and when the angle θ is allowed to range between 0 and π. Eq. 3.209 becomes:

$$F = v \int_0^{2\pi} d\phi \int_0^{\pi} r^2 \sin\theta d\theta = 4\pi r^2 v \qquad (3.210)$$

What if the source of water is not in the exact center of the sphere? In this case, the water velocity field $\vec{v} = \vec{v}(x, y, z)$ will be a function that depends on the three Cartesian coordinates x, y and z. However, since we assume that the fluid is incompressible, so much water is output by the source, so much must exit out of the closed surface providing that the source is contained inside the closed surface S. So we conclude, without having to calculate a difficult integral, that no matter where *inside* the sphere the source is located, the total integrated flux output must be the same. Obviously, when the source is shifted from the center, the vector $\vec{v}$ will not be constant at the surface of the sphere.

There is a special *class* of vector fields for which the flux has remarkable properties. Suppose the vector field depends on the inverse square of the distance. In this case, eq. 3.209 gives a flux that is always constant, independent from the radius of the sphere. This happens because as the area grows as r^2, the vector field decreases as r^2 so the product is constant.

Let's now turn to gravitation. We have seen that the gravitational field is expressed in eq. 3.203. Let's insert the gravitational field $\vec{g}(r)$ instead of $\vec{v}(r)$ in eq. 3.209. We have:

$$F = \oint_S (\vec{g} \cdot \vec{n})dA = -\oint_S \frac{GM}{r^2}\vec{n} \cdot \hat{r}\, dA \qquad (3.211)$$

In the case of a field depending only on r, we know that the vector $\vec{n}$ and $\hat{r}$ are parallel. Using eq. 3.210 with g instead of v we have:

$$F = \oint_S (\vec{g} \cdot \vec{n})dA = -\frac{GM}{r^2}4\pi r^2 = -4\pi GM \qquad (3.212)$$

We therefore obtain the so-called *Gauss's law for gravitation*[28]:

$$F = \oint_S (\vec{g} \cdot \vec{n})dA = -4\pi GM_{enc} \qquad (3.213)$$

---

[28]There is an equivalent, and more used Gauss's Law in electrostatics relating the electric field vector $\vec{E}$ to the enclosed charge q_{enc}. In the vacuum we have $F = \oint_S (\vec{E} \cdot \vec{n})dA = \frac{q_{enc}}{\epsilon_0}$, where ϵ_0 is the vacuum permittivity.

where we explicitly indicate with M_{enc} the mass contained inside the closed surface S. This kind of surface is usually called a Gauss surface.

Proving Newton's shell theorem is now relatively straightforward and is a direct consequence of eq. 3.213. A homogeneous spherical shell or a spherically symmetric mass both have an important property: the vector $\overrightarrow{n}$ is directed towards the point of symmetry. This means that the gravitational field of the shell or the spherical mass is directed exactly to the center of symmetry exactly as if all the mass was concentrated in that point. The field is radially symmetric $\overrightarrow{g} = g(r)\overrightarrow{n}$ where $\overrightarrow{n}$ points towards the center. We can then use eq. 3.213 to find what $g(r)$ we obtain for a radially symmetric shell or sphere:

$$\overrightarrow{g} = g(r)\overrightarrow{n}$$
$$\oint_S (\overrightarrow{g} \cdot \overrightarrow{n})dA = g(r)\oint_S dA = 4\pi r^2$$
$$g(r)4\pi r^2 = -4\pi G M_{enc} \tag{3.214}$$
$$g(r) = -\frac{GM}{r^2}$$

The last equation in eq. 3.214 proves Newton's shell theorem.

3.5 PLANET'S MOTION USING EULER-LAGRANGE EQUA- TIONS

We conclude this chapter with showing how to obtain the equation of motion of a planet (eq. 3.120) using the Euler-Lagrange equation 3.11 reported here for convenience in its general form:

$$\frac{\partial \mathcal{L}}{\partial q} - \frac{d}{dt}\frac{\partial \mathcal{L}}{\partial \dot{q}} = 0 \tag{3.215}$$

where the coordinates q and $\dot{q}$ are generic coordinates, not necessarily Cartesian. We have seen that we define the Lagrangian $\mathcal{L}$ as a **scalar** quantity obtained by subtracting the potential energy to the kinetic energy of the system, $\mathcal{L} = T - V$. We have already calculated the potential and kinetic energy of a planet of mass m orbiting the sun of mass M (see eq. 3.22). We now choose to use polar coordinates:

$$T = \frac{1}{2}m(\dot{r}^2 + r^2\dot{\theta}^2)$$
$$V = -\frac{GMm}{r} \tag{3.216}$$

The Lagrangian can then be written as:

$$\mathcal{L} = T - V = \frac{1}{2}m(\dot{r}^2 + r^2\dot{\theta}^2) + \frac{GMm}{r} \tag{3.217}$$

We now show how to quickly obtain eq. 3.120, whose solution gives the orbit of a planet in the solar system.

Let us now apply Euler-Lagrange equations with respect to the two coordinates r and θ. We have:

$$\frac{\partial \mathcal{L}}{\partial \theta} - \frac{d}{dt}\frac{\partial \mathcal{L}}{\partial \dot{\theta}} = 0$$
$$\frac{\partial \mathcal{L}}{\partial r} - \frac{d}{dt}\frac{\partial \mathcal{L}}{\partial \dot{r}} = 0 \tag{3.218}$$

The Lagrangian of eq. 3.216 does not have an explicit dependence on θ and we have seen that this implies the conservation of the associated momentum. The first equation in 3.218 gives:

$$\frac{\partial \mathcal{L}}{\partial \dot{\theta}} = mr^2\dot{\theta} = L = \text{const.} \tag{3.219}$$

The second equation in 3.218 gives:

$$m\ddot{r} - mr\dot{\theta}^2 = -\frac{GMm}{r^2} \tag{3.220}$$

Using eq. 3.219 and $h = \frac{L}{m}$ we finally have:

$$m(\ddot{r} - \frac{h^2}{r^3}) = -\frac{GMm}{r^2} \tag{3.221}$$

which is exactly eq. 3.120.

Think About It...

Looking at fig. 3.37 by Newton, we can certainly state that Newton was the first to think about the possibility of sending satellites in orbit around the Earth.

FURTHER READING

Doran, C., and Lasenby, A. (2003), *Geometric Algebra for Physicists*. Cambridge University Press.

Feynman, R.P., Leighton, R.B, and Sands, M.L. (1963), *The Feynman Lectures on Physics. Vol. II*. Pearson/Addison-Wesley.

Goldstein, H. (2002), *Classical Mechanics*. Addison-Wesley.

Neuenschwander, D.E. (2017), *Emmy Noether's Wonderful Theorem*. Johns Hopkins University Press.

Rojo, A., and Bloch, A. (2018), *The Principle of Least Action.* Cambridge University Press.

A Few Facts about the Solar System

CONTENTS

W E spent considerable effort in the previous chapters to calculate the functional form of the orbit of a planet around the Sun and various different methods to prove it. We have seen that Kepler's laws are derived from Newton's laws. In addition, we have seen that if the gravitational force is central, the planet/Sun system conserves the angular momentum. If the central force has the special dependence of the inverse square of the distance from the planet to the Sun, then the orbit is a conic section. The energy discriminates between closed orbits (circle and ellipses) or open orbits (parabola and hyperbola).

We now explore briefly how the motion of planets has been crucial in determining the law of gravitation. We need to go back in time and study how scientists have worked out that the Sun is at the center of the solar system and that the gravitational attraction is universal and is proportional to the inverse square of the distance between the masses.

4.1 GEOCENTRIC VERSUS HELIOCENTRIC

Let us now follow the reasoning of an ancient philosopher willing to build a model that explains all the observations above.

The first natural assumption, first dictated by the fact that the sky effectively rotates east to west regularly, was to place the Earth at the center of the Universe. This vantage position was justified also by religious assumptions about the special place that humans also have in the cosmos. If humans are

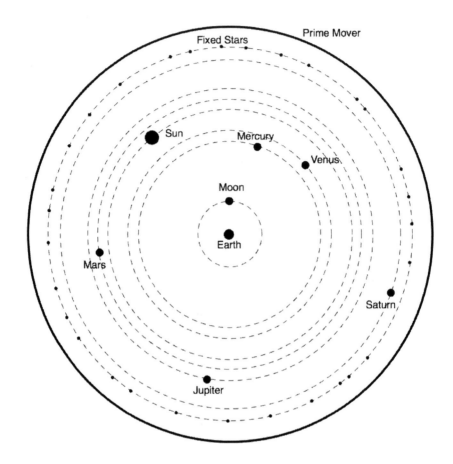

FIGURE 4.1 Aristotle's model.

central, then the Earth where we live must be special and so it does not look unreasonable to put it at the center, as well.

Aristotle (384 - 322 BC) was the most influential philosopher to put forward this geocentric view of the cosmos. He was so influential that his model survived 18 centuries before being superseded by the work of Copernicus around the 15th century. Aristotle demanded few strict conditions for the motion of objects in the sky: since they appear to move in circles and circles are perfect geometric figures, the substance making up the stars and planet must also be perfect and their shape must be spherical. So the Earth and the planets must be spherical[1].

---

[1] Aristotle supported his proposal of a spherical Earth with observations. He noticed that during a lunar eclipse, the Earth's shadow is circular and this can be only explained by a spherical Earth.

FIGURE 4.2 Aristotle (384—322 BC) was a very influential Greek philosopher. His views on astronomy survived several centuries, until the Renaissance.

Since the sphere is the perfect geometrical figure with maximum symmetry, then its center is the natural place to position the Earth. All the astronomical objects are fixed to rotating transparent spheres made of an ethereal, transparent fifth element (quintessence) to be added to the canonical four elements fire, air, earth, and water. Aristotle noticed that the relative position of all the stars did not change and therefore they must be fixed to the same sphere – the sphere of fixed stars – which is behind all other objects (see fig. 4.1). Starting from the Earth in the center, the Moon is embedded into the first sphere. This sphere is still quite close to the Earth and therefore is subjected to its influence: this explains why the Moon has phases and dark spots. We then have the spheres for Venus, Mercury, the Sun, Mars, Jupiter and Saturn and fixed stars. All these spheres rotate at constant different angular velocities and the motion is generated by an additional more external sphere which is "unmoved". This geocentric model explains pretty well why stars appear not to move with respect to each other. In fact, constellations preserved their shape with time. The apparent stable luminosity of Venus was also believed to be another fact in favor of geocentrism. Stable luminosity means that Venus is always at the same distance from Earth.

Aristotle needed 55 spheres to accommodate the astronomical data at his disposal. He spent a large amount of effort in perfecting his system so as to have a good agreement between theory and observations. In this respect he is quite modern. As observational data became more accurate and more abundant, Aristotle's model needed to be updated. The most important facts were the apparent change in luminosity of Mercury, Mars and Jupiter during the year and the *retrograde motion* (see fig. 4.3). It was in fact noticed that Mars and Jupiter were showing a retrograde motion, a phenomenon where they would seem to slow down, go backwards, and then move forwards against the fixed stars.

As measurements became more abundant and accurate, the spherical approach needed to be modified while still maintaining the philosophical requirements of "perfection" of circular motion. Ptolemy (around AD 140), a Roman citizen of Greek origin living in Alexandria, needed to introduce few modifications to concentric spherical models by introducing a combination of circular motions (epicycles) imposed on other circular motions (deferent circles). This complication was absolutely needed because it was evident that the motion of the Sun, the Moon, and the planets as seen from the Earth was not simply circular. He also removed the Earth from the center of the spheres to allow for planets and other objects to vary their distance to the Earth. This modification immediately explained the variations of luminosity of some planets. The introduction of the epicycles qualitatively explained the retrograde motion.

Therefore, according to Ptolemy, the Earth is the center of the Universe but occupies a shifted position with respect to the center of the celestial spheres as shown in fig. 4.4. In the opposite position with respect to this center, there is a point called *equant* from which the planets appear to move at constant angular speed. This geometry accounted for velocity variations and retrograde

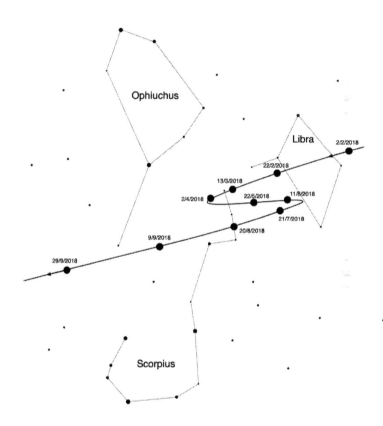

FIGURE 4.3 Retrograde motion of Mars in summer 2018. Libra constellation is at $RA = 15^h$, $\delta = -15°$.

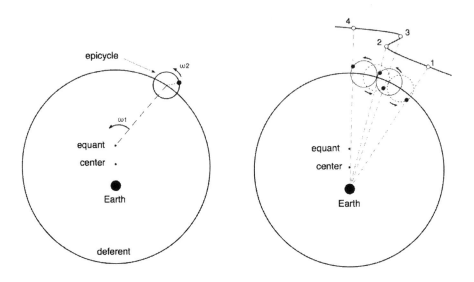

FIGURE 4.4 Ptolemy's deferent and epicycle circles.

motion of planets. It was quite complicated but, with refinements, was able to explain the observed motions. Ptolemy was happy with his system because it was capable of explaining most of the observations. But, most importantly, was giving the Earth a central position in the Universe and this was a basic requirement for any scientific theory at those times because of religious compatibility.

In fig. 4.3 we see the apparent motion of Mars against the fixed constellations of Libra, Scorpius, and Ophiuchus while in fig. 4.4 we see the solution proposed by Ptolemy. The right panel of fig. 4.4 shows 4 sequential positions of the planet moving along the epicycle as seen from Earth. If $\omega 1 < \omega 2$, i.e. if the angular velocity of the radius vector joining the center of the epicycle is less than the angular velocity of the radius describing the epicycle, then when the planet is describing the arc contained within the deferent from point 2 to point 3, it appears to be moving in a retrograde motion. Ingenious but not good enough. In fact, although qualitatively capable of explaining retrograde motion, epicycles were more and more difficult to adapt to new and more accurate data.

Already in ancient times, some philosopher attempted to propose new models where the Earth was not at the center of the Universe. Aristarchus of Samos (see chapter 1) was among the first to propose that the Sun, instead of the Earth, was in the center. Although basically correct, his theory did not have much success, probably because of the highly influential figure of Aristotle. We need to wait a couple of millennia before the times are ripe for a revolutionary proposal. Copernicus is the figure that more than others had the stature and charisma to propose a heliocentric model.

FIGURE 4.5   Ptolemy.

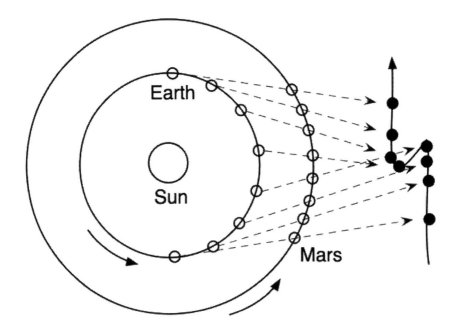

FIGURE 4.6 Retrograde motion of Mars in Copernicus's heliocentric model.

Many of the observed phenomena that we described above can be explained in a simpler way by the heliocentric hypothesis. For example, stars rise east and set west on a very regular period of 24 hours rotating around the star Polaris. It is not the sky that rotates, with all its complex structure. Instead, it is much simpler to assume that the Earth is not fixed but is free to move. Having freed the Earth, we can now accept that it can spin around its axis and the axis is pointing towards the star Polaris.

The retrograde motion was explained very easily as shown in fig. 4.6. But above all, all the complications of several spheres with epicycles were not needed anymore. Copernicus indicated the right way and, even though his model was not accurate, he paved the road for other scientists to refine his heliocentric model. Copernicus wrote his proposal in a hand-written book that he initially distributed just to his friends. In this book, in addition to placing the Sun in the center of the Universe, he attributed the Sun's rising and setting to the Earth rotating around its axis. The apparent circular motion of the stars was also explained by this rotation. The revolution of the Earth around the Sun also accounted for the seasons. He did not publish his book, *De Revolutionibus Orbium Coelestium*, until 1543 just two months before his death.

It was Kepler who finally corrected Copernicus's wrong assumption of circular orbits with elliptical orbits. When Galileo took Copernicus's ideas seriously to restate in 1632 that the Sun is at the center of the Universe, the church condemned him to house arrest for heresy and subjected him to inquisition.

4.2   MOTION AND COORDINATES

Before delving into the maths and physics, let's orient ourselves and observe the sky, paying attention to the various phenomena. In the western part of the world we enjoyed an amazing number of ancient philosophers, mostly Greek, that had the opportunity and the geniality to ponder over fundamental questions. The majesty of the night sky has certainly provided the ground for speculations about not only the origin of the world but also the mechanism responsible for the observed regularity of motion of stars and planets.

The are some observations that have sparked a number of debates among the ancient philosophers. We can try to get a feeling for what the Universe looked like at those times by pretending we know nothing about the solar system and trying to see what can we deduce by simply observing the sky. The first thing we observe is that, at night and if it is clear, there are lots of point-like objects called stars. In fig. 4.7 various important lines are shown. Suppose we are located at mid-latitude in the Northern Hemisphere and observe the sky. The big sphere around us is the apparent sphere where all sky objects seem to be located like, for example, the stars, the planets, and the Sun. If we look straight up, we define our local zenith. If we project our local horizon towards the celestial sphere, we define a circle also called the *horizon*. Obviously every

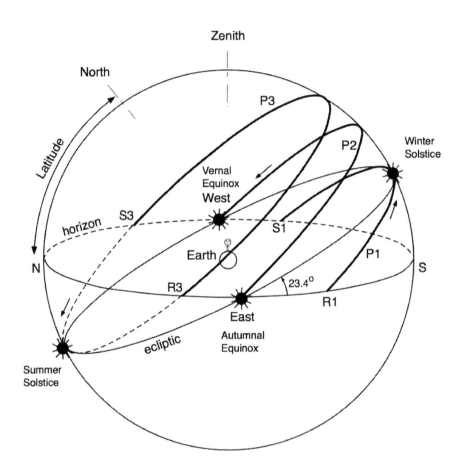

FIGURE 4.7 Motion of the Sun in the sky as seen by an observer located on Earth at mid-latitude in the Northern Hemisphere.

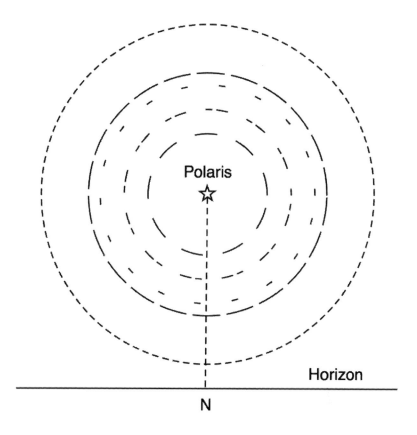

FIGURE 4.8 How to determine due north at mid-latitude in the Northern Hemisphere. Dotted lines show the traces of various stars.

observer has its own zenith and its own horizon and if we want to compare what we see in the sky we need to define a reference system common to all observers. Experience has demonstrated that cardinal points are very useful when orienting and traveling. So, let's determine the cardinal points using the motion of sky objects.

After looking at the stars for some time we notice immediately that the whole sphere seems to rotate around a fixed point very close to a star called *Polaris* (see fig. 4.8). If we bring down a line[2] perpendicular to the horizon, where this line intercepts the horizon, that is (*almost*) due north. Having found north, it is now easy to find east, south and west by turning 90° CCW in sequence. As a bonus, if we measure the angle between the Polaris and north, that angle is equal to the latitude of the observer as shown in fig. 4.7.

---

[2]In reality we need to identify the great circle passing through Polaris: North is where the great circle intercepts the horizon.

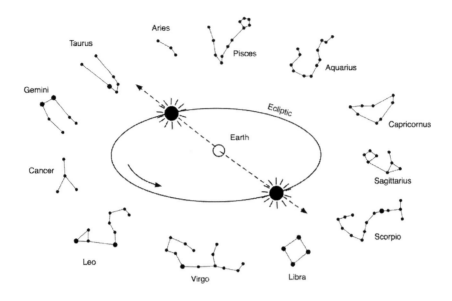

FIGURE 4.9 Apparent motion of the Sun with respect to the background of fixed stars.

Now we turn our attention to the motion of the Sun. First, we notice that the Sun culminates, i.e. reaches the highest point in the sky, always towards south while the rising and setting points R and S depend on the time of the year. In summer, the day is longer than the night and the Sun culminates very high in the sky. In winter the exact opposite happens: nights are longer than days and the Sun culminates low in the sky. There are two special days in spring and in autumn when the length of the day is equal to the length of the night. These days are termed as **equinox** from the Latin *equi*, or "equal" and *nox*, or "night". The other remarkable observation consists of the fact that the path of the Sun in the sky during the winter season $(P1)$ is lower than the path $(P3)$ during the summer as observed from our mid-latitude place in the Northern Hemisphere. There are two special days during which the Sun reaches the maximum and minimum altitude in the sky. The one in the summer is called **summer solstice**[3] (highest altitude of the Sun), while the one in the winter is called **winter solstice** (lowest altitude of the Sun). The duration of the daytime hours is minimum at the winter solstice and maximum at the summer solstice.

If we have time to spare, and apparently the ancient philosophers had

[3] From the Latin *solstitium*, the point at which the Sun seems to stand still. *Sol*, or "Sun", and *sistere*, or "stand still".

plenty, we can trace the motion of the Sun with respect to the fixed stars over a long period of time. We cannot obviously see the Sun and the background stars because of the strong sunlight. But imagine for a moment that you can do that: we immediately notice that the Sun moves eastward with respect to the background stars at *almost* the same speed as the fixed stars behind. The Sun moves and comes back in the exact same position with respect to the fixed stars after 1 year or 365 days. Well, not exactly: after 365.2422... days. This means that, with respect to the fixed stars, because one complete turn is $360°$, the Sun moves at a speed of slightly less than $1°$ per day eastward. With a bit of patience, we can trace the full path of the Sun with respect to the fixed stars and this line is called **ecliptic**[4]. As seen from Earth, the Sun moves along the ecliptic line and, depending on the month of the year, it crosses 12 famous constellations contained in the Zodiac. The Zodiac is a band in the sky centered on the ecliptic and about $16°$ wide. We notice another remarkable fact: the Moon and the visible planets are always contained within this band.

With even more patience we notice another interesting fact: the Sun seems to be moving along the ecliptic at a slightly faster speed during the summer months than the winter months.

These simple observations are hinting at something extremely interesting: the observations are consistent with having the Sun, the Earth, and the other planets including the Moon, orbiting more or less in the same plane! That is quite remarkable and hints at something very profound about the origin of the solar system.

Let's look at the Moon: it is spherical, rotates around us always showing the same face, and it goes through phases, i.e. the portions of the face directly illuminated. These portions change with time going from no illumination (new Moon) to full illumination (full Moon) and back to no illumination in about 29.53 days (*synodic month*). The synodic month is not stable but varies during the year. If we record how much time is needed for the Moon to rotate around the background of fixed stars we find that the cycle is about 27.32 days (*sidereal month*).

In addition to the obvious motion of the Sun and the Moon, there are other point-like objects (planets[5]) that also move with respect to the background of fixed stars. The ancients cataloged Mercury, Venus, Mars, Jupiter, and Saturn as planets. To complicate things a bit more, some of these planets experience strange patterns by showing *retrograde* motion, i.e. the path against the fixed stars shows loops with changes of direction between the **direct** motion and the motion in the opposite direction or **retrograde**. In fig. 4.3, the motion of Mars is represented against the fixed constellations of Libra, Ophiuchus and Scorpius. From early February to early April, Mars seems to move in the general direction from east (to the right) to west (to the left). Then from

---

[4]The word *ecliptic* comes from the Greek word εκλιπτικος "of an eclipse". The ancients in fact noticed that eclipses happened when the Moon was close to the ecliptic line.

[5]In ancient Greek, the word πλανητης means "wanderer".

early April to mid-May Mars seems to drift from west to east then reversing direction again back to the west.

All these observations naturally lead to considering the Earth at the center of a rotating Universe. We have seen in the previous sections some history of how this assumption was challenged. Let us now discuss how we can identify stars in the sky and how we can tell another observer how to find a specific star. Suppose that during a very nice and clear night we see a bright star in the sky and we want to understand what star it is. We open an astronomy book and in it we find that stars are identified by two coordinates: right ascension (RA) and declination (Dec). The problem is that if I use two coordinates that are easy for me to determine like, for example, the azimuth and the elevation of a star, these numbers change constantly with time. So it seems like I need 3 numbers to identify a star, the azimuth, the elevation and the time at which these two numbers were recorded. Another observer in, say London, will find it completely useless to know that star A had a certain azimuth and a certain elevation in Rome (so we need also to add the local coordinates of Rome, longitude and latitude). The problem is, how can we uniquely identify a star in the sky in such a way that each observer can point his/her telescope to it being sure that they are looking at exactly the same star? The first solution that comes to mind, and it is a good solution, would be to express the position of the stars on a coordinate system where the stars are fixed. In other words, we express the coordinates of stars on a rotating coordinate system. It is natural to assume as fixed rotating axis, the axis that goes through the North Celestial Pole, which coincides with the Earth's rotation axis. If we now project the Earth's equator in the sky, we have a great circle called the *celestial equator* (see fig. 4.10). We define this circle as the line of declination equal to zero. All objects on this circle will have declination = 0 coordinate.

Now that we have a reference system where the stars do not move[6], we want to know how to calculate the local azimuth (A) and elevation (E) of a star given its α and δ. In order to proceed with this calculation we need to know with some accuracy the local time (LT) and the coordinates of where our telescope is located, i.e. the latitude λ and the longitude θ.

It is useful to look at the perspective of an observer looking at the sky in the Northern Hemisphere as in fig. 4.11. The great circle going through south is the meridian. The celestial equator and the ecliptic intersects in the point γ, also called *first point of Aries* or *vernal equinox*. This point is the zero point of right ascension. There is also another intersection point, called *first point of Libra* located exactly at 180° which does not have any special meaning.

In order to find the azimuth and elevation of a star at a certain location and at a certain time, we need to make a few calculations. The first calcula-

---

[6]In reality all stars move also with respect to this reference but we neglect their motion because it is very small given their distance to Earth.

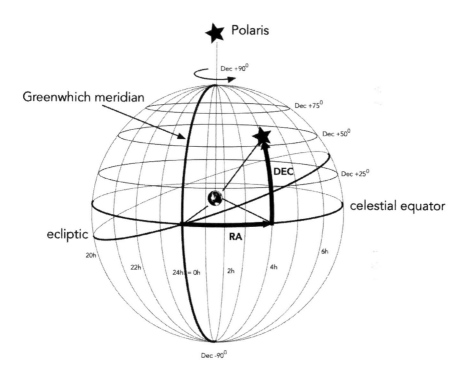

FIGURE 4.10 Right ascension (RA, α) and declination (Dec, δ) of a star. Right ascension zero is defined as the meridian passing through the special point obtained by intersecting the celestial equator and the meridian passing at the Greenwich observatory in England.

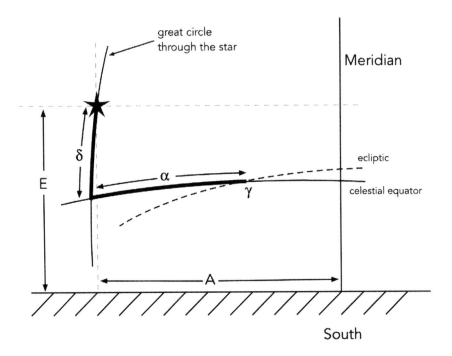

FIGURE 4.11 Celestial coordinates of a star as seen from an observer on Earth in the Northern Hemisphere.

tion consists of finding the **Hour Angle** H from the right ascension α. The conversion formula is:

$$H = LST - \alpha \qquad (4.1)$$

where LST is the *Local Sidereal Time*. The observer knows his/her local time as given by his/her watch. Therefore we need to transform the local time (LT) into local sidereal time (LST).

In everyday life we are used to knowing our local time, i.e. the time given by our watch. Local time, by convention, is measured with respect to the Sun: we take the time between two successive crossings of the Sun at the meridian and this defines 24 hours. With some degree of approximation, 12:00 noon *Universal Time* (UT) is the time when the Sun crosses the meridian at Greenwich. However, unfortunately there is a problem: keeping the time with the Sun in this way is not very accurate. The Earth spins around its axis in a very regular way, but its orbit around the Sun is not exactly circular as we have seen in the previous chapter, but is slightly elliptical. In addition, the spin axis of the Earth is inclined by 23.5° with respect to the plane of the orbit. These effects make the time kept with the Sun, for example through

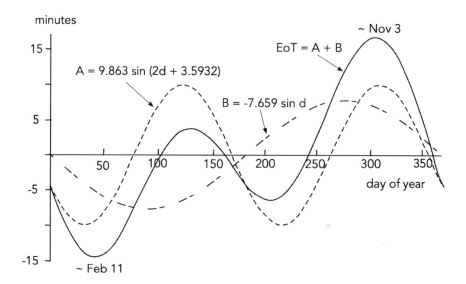

FIGURE 4.12 The Equation of Time plot. The EoT line represents the deviation in minutes with respect to the day of the year of a sundial clock on Earth with respect to a sundial orbiting around the Sun on a circular orbit and spinning around an axis perpendicular to the plane of the orbit. The EoT is the sum of the two terms A and B taking into account, respectively, the elliptical orbit and the tilt of the Earth's spin axis.

a sundial, run faster or slower depending on the position of the Earth on its orbit around the Sun. The difference between a clock running on a circular orbit spinning perpendicularly to the orbital plane (uniform clock) and the sundial clock can be as high as several minutes. The time kept by this uniform clock is referred to as *Universal Time* (UT). On Feb. 12th, the Sun can be behind by as much as 14 min and 6 sec, while around Nov. 3rd the Sun can be ahead by as much as 16 min and 33 sec.

Historically the difference between UT and a sundial is plotted into a so-called Equation of Time [7] (EoT), which is needed to correct the sundial time. It can be shown (see later) that a good approximation to the EoT is given by:

$$\Delta T = A + B = -7.659 \sin d + 9.863 \sin(2d + 3.5932) \qquad (4.2)$$

where d is the day of the year. The first term A in eq. 4.2 is due to the effect of the elliptical orbit while the second term is due the Earth's tilt.

A good watch therefore will keep the Universal Time UT corrected for the longitude. Astronomers, on the other hand, have a lot of difficulties using UT when studying the motion of stars. The reason is that there is a discrepancy between the time elapsed between two successive transits of the Sun at the meridian and two successive transits of a distant star at the meridian. Two successive transits of the Sun at the meridian define exactly 24 hours or 86,400 seconds. If we clock two successive transits of a distant star at the meridian, we discover that this happens faster, i.e. after 86,164.0905 seconds, or in 23 hours, 56 minutes and 4.0905 seconds. Therefore a clock running on successive transits of stars runs faster than a clock set on the successive transits of the Sun. This is a consequence of the fact that the Earth is orbiting the Sun!

In fig. 4.13 the Earth is orbiting the Sun in counterclockwise direction from the point A to the point B. At A, an observer on the surface of the Earth sees the Sun passing at its meridian at a certain time t exactly when a star is directly behind the Sun. So both the Sun and a star are transiting at the meridian when the Earth is at A. One day after, when the Earth is at point B along the orbit, the Sun will cross the meridian after one complete 360° revolution plus a little angle, while the star will have crossed the meridian a little bit earlier. So the star will transit the meridian slightly before the Sun. It means that one "star" day is shorter than one "Sun" day, thus justifying the difference reported above between solar time and sidereal[8] time.

It is possible to calculate the Local Sidereal Time, at least in an approximate way, by using eq. 4.3 [1]:

$$LST = 6.697374558 + 0.06570982441908D + H + 0.000026 \left(\frac{JD}{36525}\right)^2 \qquad (4.3)$$

---

[7]This is not intended as a proper equation, but rather a reconciliation of a difference.

[8]The word sidereal comes from the Latin *sidereus*, meaning "starry" or "of the constellations".

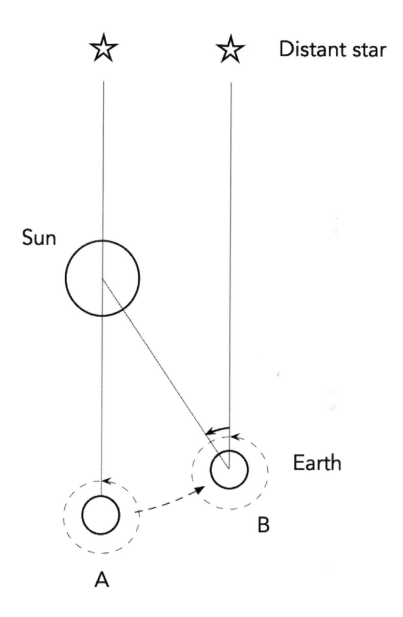

FIGURE 4.13 Sidereal time is shorter than solar time because the Earth is orbiting the Sun. At the meridian of an observer on the surface of the Earth, a distant star will transit every 23 hours, 56 minutes and 4.1 seconds rather than 24 hours. This is happening because in one day the Earth has moved from point A to point B and to have the Sun crossing the meridian there is an extra angle to be rotated.

where H is the Universal Time UT, $D = JD - 2451545.0$ and JD is the Julian Date. In the calculations involving time it is often convenient to use the so called *Julian Day* (JD). When dealing with astronomical events, very often it is important to calculate the time difference between two events. Dealing with dates expressed in year, month, day, and time of the day makes the calculation of differences not practical. Therefore astronomers prefer to use a continuous count of days starting from a Julian Day zero corresponding to noon of Monday, January 1st 4,713 BC. The Julian Date can be considered like a clock constantly advancing and identifying exactly any particular instant. The fraction of the day is added to JD, so any instant of time is identified as a integer, associated with the day, and a decimal fraction of a day, associated with the time. The Julian Date of any instant is the Julian Day number for the preceding noon in Universal Time plus the fraction of the day since that instant and this is unique all over the world, i.e. is not a local time.

Having the RA, DEC, H, and latitude (LAT) of the observing site, we can now calculate the altitude (ALT) and azimuth (AZ). We first calculate the altitude:

$$ALT = \sin^{-1}[\sin(DEC)\sin(LAT) + \cos(DEC)\cos(LAT)\cos(H)] \quad (4.4)$$

We now calculate the quantity A:

$$A = \cos^{-1}[\frac{\sin(DEC) - \sin(ALT)\sin(LAT)}{\cos(ALT)\cos(LAT)}] \quad (4.5)$$

Finally, if $\sin(H) < 0$ then $AZ = A$, otherwise $AZ = (2\pi - A)$.

Today, nobody manually calculates these quantities. Several web sites as well as computer programs will do this calculation accurately and rapidly.

4.3 THE ANALEMMA

Let's now take a picture of the position of the Sun every day at the same time – let's say noon – and see what kind of graph we obtain. It is certainly fun to do this with a real camera, but if we don't want to wait a full year we can use a website[9] to download the calculated positions of the Sun. The result is shown in fig. 4.14 where an oddly shaped figure-of-eight curve is shown. This curve is called the **analemma** and it reveals very interesting features. We notice immediately that the Sun reaches its maximum and minimum altitude respectively on the 21st of June and the 22nd of December. These two dates are respectively the summer and winter solstices. If we now trace a horizontal line exactly in the middle of these two solstice lines, we intercept the analemma at two remarkable points: the two equinoxes (vernal and autumnal). This special altitude is exactly equal to 90° minus the latitude of the observer which, in the case of Manchester (UK) is about 53.5°. We notice also that the two

---

[9]See, for example, https://ssd.jpl.nasa.gov/horizons.cgi.

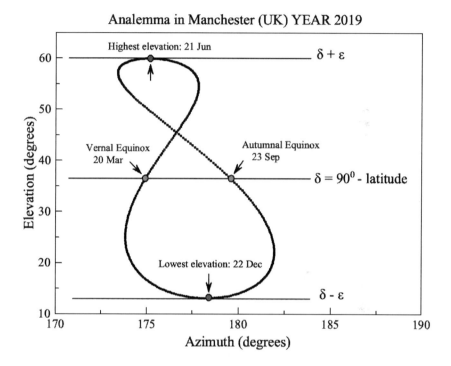

FIGURE 4.14 The analemma for the city of Manchester (UK) at longitude = 2.2° W and latitude = 53.5° N. Each black dot represents the
recorded azimuth and elevation of the Sun at 12 noon every day for
the year 2019. Notice that the two equinoxes happen at an altitude of
$\delta = 90°$ minus the latitude of Manchester. The two solstices happen
at altitudes of respectively $\delta + \epsilon$ and $\delta - \epsilon$, where ϵ is the tilt of the
Earth's axis with respect to the plane of the orbit around the Sun. The
asymmetry is explained in the text.

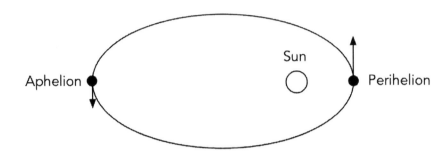

FIGURE 4.15 Perihelion and aphelion points along the Earth's orbit.

altitudes, maximum and minimum, can be obtained by adding to δ the special angle $\epsilon = 23.4°$. We will see shortly the significance of this angle.

The analemma in fig. 4.14 shows an evident tilted asymmetrical figure-of-eight shape having the lower loop bigger than the upper.

The analemma is due to the combination of two causes: the eccentricity of the Earth's elliptical orbit and the tilt of the Earth's spin axis with respect to the plane of the orbit [8]. These two effects have different periodicity: the eccentricity generates an effect proportional to a period of 365 days (annual) while the Earth's tilt generates an effect with half the period (semi-annual), or twice the frequency. These two periodicities are connected with the two periods shown in fig. 4.12 thus hinting at a close relationship between the Equation of Time and the analemma. We now show that this is indeed the case and we justify the functional form of eq. 4.2.

In this demonstration we follow closely the reference [8]. We want to find the variations of the periodicities of the motions of the Earth when spinning around its axis and when revolving around the Sun, with respect to the ideal motion of the Earth revolving around the Sun in a perfect circular orbit and with its axis perpendicular to the plane of the orbit. We will see that these deviations are responsible for the Equation of Time and analemma features. We start with the definition of angular momentum of the Earth as expressed in eq. 3.24. If we consider the angular momentum *per unit mass*, we can write $h = r^2\dot{\theta}$ and we know that this quantity is constant during the orbital motion. Let's consider the two special points *perihelion* and *aphelion* which are, respectively, the closest and farthest points of the Earth's orbit around the Sun (see fig. 4.15).

Using equations 3.76 and 3.77, the distances from the Sun to the Earth at perihelion and aphelion can be written as:

$$r_P = (1 - e)a$$
$$r_A = (1 + e)a$$
(4.6)

where we used the conditions of $\theta = 0$ and $\theta = \pi$ respectively for the perihelion and the aphelion. In fig. 4.15 it is also evident that, for an elliptical orbit, the angular velocities $\omega_P = \dot{\theta}_P$ and $\omega_A = \dot{\theta}_A$ will be different. Kepler's laws tell us that the planet will be faster at perihelion and slower at aphelion in order to keep constant the areas swept by the radius vector per unit time. We can therefore write:

$$h = r^2\dot{\theta} = (1 - e)^2 a^2 \omega_P = (1 + e)^2 a^2 \omega_A = const. \tag{4.7}$$

where the Earth's orbital eccentricity is $e = 0.0167086$. From eq. 4.7 we notice that, during one orbit, the angular velocity varies between a minimum value ω_A and a maximum value ω_P given by:

$$
\begin{aligned}
\omega_A &= \frac{h}{a^2} \frac{1}{(1-e)^2} \\
\omega_P &= \frac{h}{a^2} \frac{1}{(1+e)^2}
\end{aligned} \tag{4.8}
$$

We now use the fact that $e \ll 1$ to approximate eq. 4.8:

$$
\begin{aligned}
\omega_A &= \frac{h}{a^2} \frac{1}{(1-e)^2} \approx \frac{h}{a^2}(1 + 2e) \\
\omega_P &= \frac{h}{a^2} \frac{1}{(1+e)^2} \approx \frac{h}{a^2}(1 - 2e)
\end{aligned} \tag{4.9}
$$

where we used only the first-order term of the series expansion for $\frac{1}{(1\pm e)^2} \approx (1 \mp 2e)$. We can now estimate an average value for the angular velocity $\bar{\omega}$:

$$\bar{\omega} \approx \frac{(\omega_A + \omega_P)}{2} = \frac{h}{a^2} \tag{4.10}$$

and eq. 4.9 becomes:

$$
\begin{aligned}
\omega_A &= \bar{\omega}(1 + 2e) \\
\omega_P &= \bar{\omega}(1 - 2e)
\end{aligned} \tag{4.11}
$$

If we call Ω the very stable Earth's rotation angular velocity, an inspection of fig. 4.13 shows that the apparent rotation of the Sun, as seen from Earth, happens with angular velocity $\Omega - \omega$. In fact, the rotation of the Earth Ω compounds with the revolution of the Earth ω to make the crossing of the Sun at the meridian happen later than the same crossing of a distant star. Now we want to estimate the maximum fractional change:

$$\Delta = \frac{(\Omega - \omega_A) - (\Omega - \omega_B)}{\Omega} = \frac{\omega_B - \omega_A}{\Omega} \approx \frac{4e\bar{\omega}}{\Omega} \tag{4.12}$$

The quantity Δ represents the maximum variation of the solar day and can be calculated to be of the order of $\pm\frac{2e\bar{\omega}}{\Omega} \approx 0.00009$ seconds per second. In a day this deviation amounts to about 7.6 seconds. The functional form of the length of the solar day due to the eccentricity of the Earth's orbit can then be expressed by a co-sinusoidal law with amplitude $\pm\frac{2e\bar{\omega}}{\Omega}$ according to:

$$\Delta_{ecc} = \frac{2e\bar{\omega}}{\Omega} \cos\left(\frac{2\pi d}{365}\right) \tag{4.13}$$

where d is the number of days elapsed since the perihelion. Eq. 4.13 explains the first term in the Equation of Time 4.2. The explanation of the effect of the Earth's tilt, or *obliquity*, is an exercise of spherical trigonometry and we leave it to the interested reader (see [8]).

4.4   TIDES IN THE SOLAR SYSTEM

The Sun is certainly the most prominent object in the sky and we will dedicate a chapter later on about some interesting physics about it. After the Sun, the Moon is the next most prominent object. We will study now one of the most important effects that the Moon exerts over the Earth: the tides. This important phenomenon has captivated philosophers for many years mainly because of the observation of the changing of the height of the sea level twice a day. One of these philosophers, Seleucus of Seleucia[10], was particularly intrigued by the regularities of the rising and falling of the sea level, what today we call tides.

Seleucus noticed many other regularities, including a yearly cycle. It seems, but it is not certain, that based on this observation Seleucus made the hypothesis that the Earth-Moon system rotates around the Sun. He certainly was the first to propose that tides are generated by the Moon.

Let's understand why we have tides and what causes them. Tides are the regular increasing and decreasing of the sea level that can be observed at the sea side. The amplitude of the variation is strongly dependent on the location. The highest tidal range, i.e. the height difference between the highest and lowest tides, is in the Bay of Fundy in Canada. In this specific coastal location the tidal range can be as high as 16.3 meters. In the United Kingdom, in the Severn estuary[11], the tidal range can reach 15 meters. In open sea the tidal range is of the order of 0.6 meter and therefore some sort of amplification must occur where the coastal line has favorable shapes. We can imagine that if a water flow is compressed or funneled, then the tidal range might be increased. The Severn estuary is clearly shaped as a funnel and there are situations for which the leading edge of the incoming tide can produce a wave capable of propagating upstream against the direction of flow of the river. This anomalous wave is called *tidal bore*.

---

[10]Somewhere near Baghdad in Iraq close to the west bank of the Tigris river.

[11]The estuary of the river Severn is located in the southern UK. The Severn is the longest river in the UK and its mouth becomes the Bristol channel.

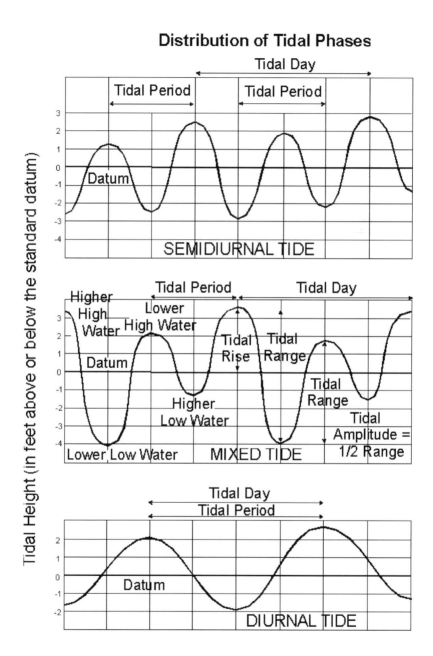

FIGURE 4.16 Diurnal and semi-diurnal tide variations. Figure adapted from https://co-ops.nos.noaa.gov/restles4.html.

In fig. 4.16 we see a plot of a typical tidal range. The top panel shows the semi-diurnal variation, i.e. two complete low-high tides per day. The bottom panel shows a rarer diurnal tide, i.e. one complete low-high tide per day. The center panel shows the mixed tide. We will see that the semi-diurnal tide is due to tidal forces associated with the Moon while the diurnal is due to interference phenomena with the lands obstructing the flow of water.

Let's see if Newtonian mechanics is capable of giving a dynamical explanation of the tides. In order to do so, we need to set up a model and we need to make a series of simplifying assumptions. In the following we will assume that the gravitational force obeys Newton's gravitational law of inverse square. We ignore the gravitational attraction of the Sun, although we will calculate later what tides are due to the Sun. We assume that the Earth is completely covered in water so as to disregard effects due to the coastal lines, and finally, we put ourselves on a reference system where also the Earth's gravitational pull is switched off: this is not difficult to imagine because artificial satellites orbiting the Earth are in this exact situation. Therefore, the Moon is the only important source of gravity and we can now study its effects at the surface of the Earth.

Let us now consider a system orbiting the Earth like, for example, the International Space Station (ISS). It is well known that the astronauts and the objects on board do not experience any gravity due to the Earth because they are constantly free falling. Being in orbit, as we have seen in previous chapters, is like falling constantly with enough horizontal velocity that you come back to the same original point after one revolution around the Earth. Now suppose that you do a little experiment inside the ISS: imagine dispersing a collection of little stones on a perfectly spherical shell like in fig. 4.17. Let us suppose that an observer is positioned in the center of such a shell at point C. An external observer will see the forces as depicted in the left panel of fig. 4.17. All the forces, and accelerations will be directed towards the center of the Earth O and since the shell has a finite size, the four masses will feel different forces in direction and/or intensity. If we call R the distance $\overline{OC}$ between C and the center of the Earth O, and r the radius of the shell, we see immediately that the force in P_1 is less than the force in P_4 because P_4 is farther away than P_1. These two forces are directed along the same direction but differ in strength.

The two masses in P_2 and P_3 are located at the same distance from O, but since they displaced by $2r$, the direction of the force is not the same and so the two forces are different in direction but of equal intensity. If we call $\vec{f_0}$ the force in C and $\vec{f_1}$, $\vec{f_2}$, $\vec{f_3}$ and $\vec{f_4}$ the forces at respectively P_1, P_2, P_3 and P_4, we can analyze the forces and their differences.

The tidal forces are the forces observed in the reference frame centered in C. This means that in order to calculate the tidal force on the points P_1, P_2, P_3 and P_4 we need to subtract the Earth force $\vec{f_0}$. First let's calculate the tidal forces t_1 and t_4 on the points P_1 and P_4. We have:

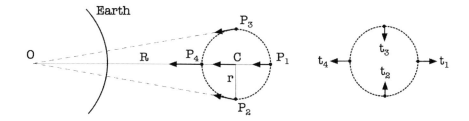

FIGURE 4.17 Tidal forces in a free-falling reference frame centered in C. The 4 non-interacting small masses P_1, P_2, P_3 and P_4 are subject to different forces. When viewed by an observer in C, tidal forces are evident as in the right panel of the figure.

$$\vec{t_4} = \vec{f_4} - \vec{f_0}$$
$$\vec{t_1} = \vec{f_1} - \vec{f_0}$$
$$(4.14)$$

All the vectors in eq. 4.14 are parallel and therefore the t vectors will be parallel to the f vectors. In addition, we immediately know that t_4 is positive because f_4 is bigger than f_0 and therefore t_4 is oriented towards the center of the Earth. This also means that the particle at P_4 will be subject to an acceleration towards the Earth and will drift away from the point C. The observer in C will see the point in P_4 drifting away from him/her in the direction of the Earth. The opposite happens for the point P_1. In this case, now the vector t_1 is negative because f_0 is bigger than f_1. This means that now the vector t_1 is oriented in the exact opposite direction away from the center of the Earth. This means that the observer in C will see that point P_1 will drift away in the opposite direction with respect to the Earth. Having determined the direction of the two vectors we can now calculate their magnitude. Using Newton's law of gravitation for the magnitudes, eq. 4.14 becomes:

$$f_4 = \frac{GMm}{(R-r)^2} - \frac{GMm}{R^2}$$
$$f_1 = \frac{GMm}{(R+r)^2} - \frac{GMm}{R^2}$$
$$(4.15)$$

where M is the mass of the Earth, m is the mass of the particles, $R = \overline{OC}$ and r is the radius of the shell. If we restrict our calculation to accelerations, we divide numerator and denominator by R^2 and remembering that $a = f/m$, then we can re-write eq. 4.15 as:

$$a_4 = \frac{GM}{R^2(1 - \frac{r}{R})^2} - \frac{GMm}{R^2}$$

$$a_1 = \frac{GM}{R^2(1 + \frac{r}{R})^2} - \frac{GMm}{R^2} \tag{4.16}$$

We now use the fact that $r \ll R$ so we can simplify eq. 4.16. Let's study the function:

$$f(\frac{r}{R}) = \frac{1}{(1 \mp \frac{r}{R})^2} \tag{4.17}$$

Having assumed that $r \ll R$ we can use the series expansion 1.42. If we set $x = \frac{r}{R}$, the first few terms of the series are:

$$f(x) \approx f(0) + f'(0)x + f''(0)\frac{x^2}{2!} + \ldots \tag{4.18}$$

where $f(0)$ is the value of the function $f(x)$ evaluated at $x = 0$, $f'(0) = \frac{df}{dx}|_{x=0}$ is the first derivative evaluated at $x = 0$, $f''(0) = \frac{d^2 f}{dx^2}|_{x=0}$ is the second derivative evaluated at $x = 0$, and so on. If we additionally neglect terms in x^2, we just need to evaluate the first two terms of the series expansion 4.18. Using the function 4.17 we have:

$$f(0) = 1$$

$$f'(0) = \pm\frac{2}{1 + x^3} = \pm 2 \tag{4.19}$$

and so we can finally write:

$$f(\frac{r}{R}) = \frac{1}{(1 \mp \frac{r}{R})^2} \approx 1 \pm \frac{2r}{R} + \ldots \tag{4.20}$$

Inserting this last expression into eq. 4.16 and simplifying, we finally obtain a good approximated formula for the intensity of the radial tidal acceleration:

$$a_4 = -\frac{2GMr}{R^3}$$

$$a_1 = +\frac{2GMr}{R^3} \tag{4.21}$$

The minus sign for a_4 indicates that the acceleration is towards the Earth while the plus sign for a_1 indicates that the acceleration is opposite. Eq. 4.21 gives the magnitude of the two tidal vectors a_4 and a_1 of the right panel in fig. 4.17.

For the two transverse components $a_2 = \frac{t_2}{m}$ and $a_3 = \frac{t_3}{m}$ we need to look

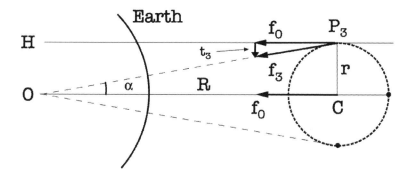

FIGURE 4.18 Transverse tidal forces in a free-falling reference frame centered in C.

at the vectors in 2 dimensions since now forces or accelerations are not along the same line.

Fig. 4.18 is a simplified version of fig. 4.17 where we have explicitly indicated the vectors needed to calculate the transverse tidal acceleration. The line $\overline{HP_3}$ is drawn parallel to the axis $\overline{OC}$. The angle $\alpha = \angle P_3OC$ is equal to the angle $\alpha = \angle HP_3O$. Omitting for simplicity for now the vector signs above the symbols, from the figure it is easy to realize that the transverse tidal acceleration is $a_3 + f_0/m = f_3/m$. From the triangle $\triangle P_3OC$ we see that $\tan \alpha \approx \alpha = \frac{r}{R}$ since $r << R$. In the same limit, we can approximate the length $\overline{OP_3} \approx \overline{OC} = R$ so we can write:

$$f_3/m \approx -\frac{GM}{R^2}$$
$$a_3 = -\frac{GM}{r^2}\sin\alpha \approx -\frac{GM}{r^2}\frac{r}{R} \tag{4.22}$$

and similar expressions for the other transverse acceleration $a_2 = -a_3$. Comparing eq. 4.22 with eq. 4.21 we see immediately that the longitudinal component of the tidal force is twice as big as the transverse component. In addition, both accelerations are proportional to R^{-3}.

We just learned that tidal forces in finite objects originate from the difference in the gravitational force between the top and the bottom part of the body as well as the different direction between the left and the right side of the body. These forces tend to elongate and stretch the object from spherical to ellipsoidal shape. In particular, tidal effects will be more noticeable with water where the intermolecular forces are not strong enough to preserve the shape. Water can easily change shape and we will see next how this is related to the phenomenon of tides.

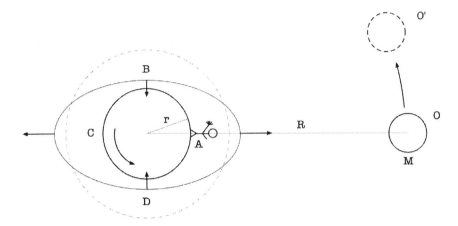

FIGURE 4.19 Top view of the Earth-Moon orbital plane. The arrows show the tidal forces on the surface of the Earth due to the gravitational attraction of the Moon. The Earth is assumed to be completely covered in water. Tidal forces due to the Moon will deform the spherical mass of water into an ellipsoidal shape. During one day, the Moon will orbit around the Earth from O to O' while the Earth would have done a complete rotation around its axis. The tidal bulge will be oriented always towards the Moon while the Earth will rotate. As a consequence, an observer on the surface of the Earth in one day will see a high tide in A, a low tide in B, a high tide in C, and a low tide in D, thus explaining the semidiurnal tide cycle.

We are now equipped to study the effect of the Moon's gravity on Earth and specifically how sea tides on Earth are generated by the Moon. Let us now assume that the Earth is free falling towards the Moon[12] or, in other words, let's see what tidal forces are present on the surface of the Earth due to the moon. Let also assume that the Earth is completely covered in water as depicted in fig. 4.19.

We know as a fact that the Moon completes one revolution around the Earth in about 27 days. In fig. 4.19 we see the Moon, of mass M, on the right orbiting the Earth counterclockwise. Since the orbital period is about 27 days, in one day the Moon will move from the point O to the point O'. During this time, the Earth would have completed one full counterclockwise rotation around its axis so the observer from A will go to the positions B, C and D

---

[12]In reality the Earth-Moon system orbits around the common center of mass and therefore a more accurate statement would be that the Earth and the Moon are free-falling towards each other.

in sequence. We assume that the Earth is completely covered in water so that we can neglect all the complexities coming from the continents impeding the flow of water.

We have seen that the tidal forces will deform a spherical shape of weakly or non-interacting bodies into an ellipsoidal shape. Since water can be easily deformed, the spherical shape of water will be changed into an ellipsoidal shape. There will be two protruding bulges in A and C and two restrictions in B and D. The ellipsoid will track the Moon during its revolution around the Earth, while the Earth will rotate around its axis. Since the Earth's rotation is much faster than the Moon's revolution, the Earth will rotate inside the ellipsoid. An observer in A will therefore see a high sea level when in A, then a low sea level when in B, then again a high sea level when in C, and finally a low sea level when in D. This explains the semi-diurnal tide cycle (see fig. 4.16.

We might wonder what would be the tidal forces on Earth due to the Sun. After all, the Sun has a huge mass compared with the Moon. Let's compare these tidal forces. The gravitational force and the tidal forces due to the Moon are given by:

$$
\begin{aligned}
F_{moon} &= \frac{GM_E M_M}{R_{EM}^2} \simeq 1.98 \cdot 10^{20}\, N \\
T_{moon} &= \frac{2GM_E M_M}{R_{EM}^3} R_E \simeq 6.6 \cdot 10^{18}\, N
\end{aligned}
\tag{4.23}
$$

where G is Newton's constant, R_E is the radius of the Earth, M_E is the mass of the Earth, M_M is the mass of the Moon, R_{EM} is the distance from the Earth to the Moon, and N is the Newton unit of force: 1 Newton of force will accelerate the mass of 1 kilogram to an acceleration of 1 meter per second per second.

Let us now calculate the same quantities for the Sun, i.e. the gravitational and tidal forces exerted on the Earth by the Sun:

$$
\begin{aligned}
F_{sun} &= \frac{GM_E M_S}{R_{ES}^2} \simeq 3.5 \cdot 10^{22}\, N \\
T_{sun} &= \frac{2GM_E M_S}{R_{ES}^3} R_E \simeq 3.0 \cdot 10^{18}\, N
\end{aligned}
\tag{4.24}
$$

where M_S is the mass of the Sun and R_{ES} is the distance from the Earth to the Sun.

Comparing eq. 4.23 with eq. 4.24 we see that while the Sun clearly and fortunately dominates the gravitational attraction, the Moon has a tidal influence on Earth roughly double that of the Sun. This means that Sun-related tides should also be visible but with less amplitude.

Looking at equations 4.23 and 4.24 we notice that the gravitational attraction of the Sun on the Earth is more than 100 times the gravitational

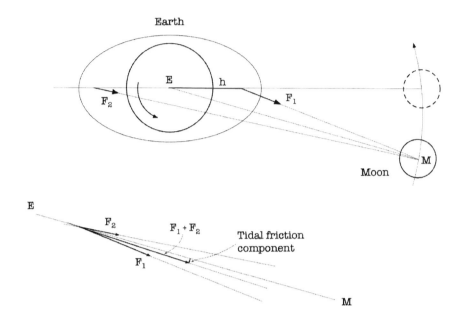

FIGURE 4.20 Top view of the Earth-Moon orbital plane. The rotation of the Earth around its spin axis pushes the tidal bulge ahead of the Moon position (top panel). The closest bulge to the Moon feels a force F_1 bigger than the force due to the far bulge on the opposite side F_2. The resultant force is not exactly aligned along the Earth-Moon axis EM having a component perpendicular to EM and directed against the rotation of the Earth (bottom panel). This component tends to slow down the Earth's rotation and push away the Moon into higher orbit.

attraction of the Moon. This is no surprise since it is the Sun that keeps all planets in orbit. However, the tidal force of the Sun is now about half of the tidal force exerted by the Moon. The R^{-3} dependence of the tidal force wins over the largest mass of the Sun. We can therefore establish that it is the Moon that dominates the observed sea tides on the Earth as properly established by many physicists including Newton.

We have mentioned earlier that the Earth's rotation period (1 day) is much faster than the Moon's rotation period (27 days). If the Earth were rotating at exactly the same rate as the Moon's rotation then the tidal bulge would always be oriented towards the Moon as shown in fig. 4.19. But we know that the Earth spins much faster and therefore we expect that the bulge is pushed ahead of the apparent position of the Moon as shown in fig. 4.20, top panel.

There are two observational evidences: 1) The Earth rotation period (the length of the day) is increasing, i.e the duration of one day is increasing with time; 2) the Moon is increasing its distance from the Earth with time. These two effects are very small but measurable. It is believed that the duration of the day was 21 hours 600 million years ago[13]. In a perfectly spherical Earth, i.e. without bulges, there is no mechanism to apply a torque capable of counteracting the rotation of the Earth. In order to be able to apply a torque, it is necessary that the body is non-spherical. In particular, if the deformation is ellipsoidal, like the case of tidal forces, then a differential pull exerted on the two bulges has a resulting component acting against the rotation of the Earth. The effect is referred to as *tidal friction* because it slows down the rotation like the friction applied by a brake.

An interesting question arises: if the Earth's rotation period is increasing, its angular velocity is decreasing with time. This means that the angular momentum is decreasing with time instead of being conserved. Where is the angular momentum going? The answer is that the Moon is acquiring the angular momentum lost by the Earth by slightly accelerating and therefore increasing the orbital radius. So, as a result, the Earth's day is slowing down and the Moon is receding from the Earth. The rate is about 4 cm per year. The non-zero torque applied to the Earth must be exactly balanced so as to have conservation of angular momentum for the Earth-Moon system. The only way to balance the angular momentum is to have the Moon increasing the orbital radius. The Moon is therefore slowly spiraling out towards larger and larger circular orbits.

Will the Moon continue to recede from Earth? Will the Earth's rotation continue to slow down until it stops? As we have seen, the Earth's rotation around its axis is slowing down due to tidal friction. There will be a time, in the far future, when the Earth's rotation spin will exactly match the Moon's rotation spin. This special condition will be evident because the Earth and the Moon will always show the same face to each other. This means that the Moon will only be visible from one side of the Earth and the length of the day will match the Moon's orbital rate. This condition is referred to as **tidal locking**.

We can try to make a simple order-of-magnitude estimation of the Earth-Moon tidal locking. We start by calculating the total angular momentum of the Earth-Moon system, considered as isolated from the influence of other solar system bodies. We have very briefly mentioned the concept of *angular momentum* defined in eq. 3.24. If we call $\omega = \dot{\theta}$ the angular velocity , i.e. for a spinning object the amount of degrees/sec of its rotation, we can rewrite eq. 3.24 as:

$$L = m\omega R^2 \qquad (4.25)$$

---

[13]There are various indications that the day was shorter in the past. By studying tidal rhythmites, i.e. layers of sediments deposited cyclically with tides, it is possible to empirically determine that the duration of the day is slowly increasing.

where m is the mass of the object rotating at a distance R around a fixed point with angular velocity ω. In isolated systems the total angular momentum is conserved. There is an equivalent conservation law for linear momentum: the total linear momentum of an isolated system is conserved. Linear momentum of a particle of mass m and velocity v is $p = mv$. There is a crucial difference between linear and angular momentum: while linear momentum does not depend on where we locate the coordinate system, the angular momentum *does* depend on where we put the origin of the coordinate system. Linear momentum contains the derivative of the position (in one dimension $v = \dot{x}$) while angular momentum contains r and $\dot{\theta}$. So angular momentum does not depend on the initial value of the angle of rotation but *does* depend on where the origin of the coordinate system is.

We need to make an important distinction: there are two types of angular momentum. There is angular momentum due to a particle of mass m orbiting around a center: this is called *orbital angular momentum*. If the body, like the Moon or the Earth is a solid body, then it can spin around an axis passing through its center of mass. The angular momentum associated with the spin motion is called *spin angular momentum*.

Before we can study the tidal locking of the Earth-Moon system, we need to express the various terms in the total angular momentum. Since the Earth is more massive that the Moon we can assume that it is the Moon revolving around the Earth instead of the more correct assumption of the Earth-Moon system orbiting around the center of mass of the system. The angular momentum of the Earth-Moon system will be the sum of the spinning angular momentum of the Earth plus the orbital angular momentum of the Moon orbiting the Earth. Neglecting other sources of angular momentum and assuming that the Earth is a solid sphere of constant density, then the total angular momentum is the sum of two terms:

$$L_0 = L_{sE} + L_{oM} = I_E \omega_E + mR^2 \omega_M \tag{4.26}$$

where I_E is the *moment of inertia* of the Earth, ω_E is the spinning angular velocity of the Earth, i.e. one rotation per day, m is the mass of the Moon, R is the distance from the Earth to the Moon and ω_M is the orbital period of the Moon which is about 28 days. The moment of inertia is the equivalent of the mass in linear momentum. It gives an indication of inertia to spin an object around an axis, for example, a symmetry axis passing through the center of mass. It depends strongly on the mass distribution inside the body. In the case of the Earth, considering that the shape is not exactly spherical and that the internal mass distribution is non-trivial, the calculation of the moment of inertia is somehow complicated. It turns out that an accepted value[14] is $I_E = 8 \cdot 10^{37}$ kg m^2/s.

---

[14]The reader might wonder where the mass M of the Earth is in this calculation. It is contained in the spinning moment of inertia which, for a homogeneous sphere spinning around an axis passing through its center is equal to $\frac{2}{5}MR^2$ where M is the mass of the Earth and R is its radius. If we use this formula using as R the Earth's mean radius and

We can calculate the total angular momentum L_0 of the Earth with the known parameters as of today. If we insert for the mass of the Moon $m = 7.35 \cdot 10^{22}$ kg, for the distance from the Earth to the Moon $R = 3.84 \cdot 10^8$ m, for the Earth spinning angular velocity $\omega_E = 2\pi/86,400 = 7.26 \cdot 10^{-5}$ rad/sec, for the Moon orbital angular momentum $\omega_M = 2\pi/(27 \cdot 86,400) = 2.7 \cdot 10^{-6}$ rad/sec, we have:

$$L_0 = I_E\omega_E + mR^2\omega_M = 5.8 \cdot 10^{33} + 2.9 \cdot 10^{34} = 3.48 \cdot 10^{34} \ m^2 kg/s \quad (4.27)$$

Notice that in eq. 4.27, although the Moon has a smaller mass than the Earth, its orbital angular momentum is 5 times larger.

We want to study under what conditions there will be a tidal locking in the Earth-Moon system. We have seen that the Earth is transferring angular momentum to the Moon, i.e. its spinning is slowing down. The angular momentum acquired by the Moon pushes it far away. There will be a time when the spinning angular velocity of the Earth exactly matches the orbital angular velocity of the Moon: let's call this angular velocity ω_L. The new distance from the Earth to the Moon when this will happen is indicated with s. Because the total angular momentum is conserved, we can write:

$$L_0 = ms^2\omega_L + I_E\omega_L \quad (4.28)$$

In order to simplify the calculation, let's neglect the second term on the right-hand side of eq. 4.28. We are somehow justified because we already noticed that today the spinning angular momentum is only 20% of the total angular momentum. In the future it will be much less. With this assumption, we have:

$$ms^2\omega_L = L_0 \quad (4.29)$$

In eq. 4.29 we have two unknowns, s and ω_L. But we can find a relationship between these two quantities by imposing that, for a body orbiting another larger body, the centripetal force must exactly balance the gravitational force. We have:

$$m\frac{v^2}{s} = \frac{GMm}{s^2}$$
$$v = \omega_L s$$
$$\frac{m\omega_L^2 s^2}{s} = \frac{GMm}{s} \quad (4.30)$$
$$\omega_L = \sqrt{\frac{GM}{s^3}}$$

---

for M its mass, we obtain a higher value for the Earth's moment of inertia of $9.7 \cdot 10^{37}$ kg m^2/s. The fact that the value calculated in this way is higher than the accepted value is an indication that the inner Earth is not homogeneous and its core density must be somehow higher.

where M and m are, respectively, the mass of the Earth and of the Moon. Inserting the expression for ω_L found in eq. 4.30 into eq. 4.29 we have:

$$s = \frac{L_0^2}{m^2 GM} \simeq 560,000 \ km \tag{4.31}$$

We can finally find the orbital period $T_L = \frac{2\pi}{\omega_L}$ by using Kepler's third law:

$$T_L = \sqrt{\frac{4\pi^2 s^3}{GM}} \simeq 48 \ days \tag{4.32}$$

4.5 ROCHE LIMIT

Tidal forces are of fundamental importance in the solar system, not only for the influence of the Moon on Earth but also for other interesting phenomena. Suppose we have a small planet of radius r and mass m orbiting a much larger planet of mass M. We would like to ask the following question: under what conditions are the tidal forces, due to the large planet, equal to the gravitational force of attraction on the surface of the small planet due to the large planet? Why is this interesting?

Let's start with the simple case of a small planet which is composed of a collection of small particles held together by just the gravitational attraction among them. We can imagine taking a large collection of small stones of total mass m, and disperse them in a spherical shape or radius r while orbiting a large planet of mass M located at distance R. We have seen that tidal forces will produce a deformation of the spherical shape into an elliptical shape but now we have to take into account that the little particles are gravitationally attracted to each other. We have two cases: if the tidal forces are larger than the gravitational force among the particles, then the spherical shape will disintegrate and will tend to form rings. If the gravitational forces are larger than the tidal forces, then the particles will not disperse and remain bound into an ellipsoidal shape.

Let's write down the tidal force F_T and the gravitational force F_G at the surface of the spherical collection of particles as shown in fig. 4.21:

$$F_G = \frac{Gm\mu}{r^2}$$
$$F_T = \frac{2GM\mu}{R^2} \frac{r}{R} \tag{4.33}$$

The Roche limit is obtained by equating the two expressions in eq. 4.33. We have:

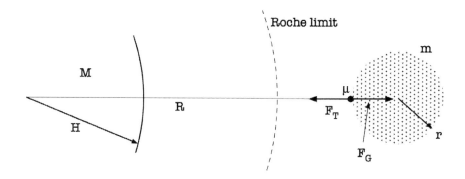

FIGURE 4.21 Roche limit for a spherical collection of gravitationally bound particles of total mass m and radius r. A small particle of mass μ at the surface and situated at a large distance $R \gg r$ will be subject to a tidal force F_T towards the large body and a gravitational force F_G towards the center of the collection of particles. The Roche limit (shown as a dashed line) is the distance at which the two forces are equal.

$$\frac{Gm\mu}{r^2} = \frac{2GM\mu}{R^2}\frac{r}{R}$$
$$R = r\left(\frac{2M}{m}\right)^{1/3} \tag{4.34}$$

The R expressed in eq. 4.34 is the so-called *Roche limit*.

We can express the Roche limit as a function of the radius H of the large body and the ratio of the densities of the large body and the small gravitationally bound body. Assuming constant densities, it is easy to show that the Roche limit can also be expressed as:

$$R = H\left(\frac{2\rho_M}{\rho_m}\right)^{1/3} \tag{4.35}$$

where ρ_M and ρ_m are, respectively, the densities of the large and the small bodies. We need to point out that the case we have studied is very simplified. In the solar system, rigid bodies are held together by strong forces and consequently the Roche limit will be different.

4.6 MEASURING THE SPEED OF LIGHT IN THE SOLAR SYSTEM

One of the major problems in ancient times was the determination of the longitude of a ship. Many lives have been lost because ships lost their position in the open sea. The main cause of death was the scurvy disease, i.e. vitamin C deficiency due to the long times at sea without fresh fruits or vegetables. It is not difficult to imagine what enormous commercial problem uncertain navigation was. A spherical coordinate system is obviously the best choice of coordinates on a sphere like our Earth. The equator main circle identifies the origin of the latitude coordinates. Such a choice is quite natural due to the rotation of the Earth with respect to the stars. The equator is the line that is exactly equidistant from the two poles. The poles are determined by the intersection of the Earth's rotation axis with the surface of the Earth. Having identified the equator, we can trace a set of parallel circles at intervals of equal angles called **latitude**. The equator is at latitude zero while the two poles are at latitude $+90°$ and $-90°$ for respectively the North and the South Poles.

If we are lost at sea, the stars provide an easy way to determine the latitude of our position. If we use the fact that the star Polaris is aligned with the Earth's rotation axis[15] we notice that it appears practically stationary in the sky with all the other stars rotating around it. If we measure the height above the horizon of Polaris, this angle is the latitude of our position. In addition, as an added bonus, Polaris also gives due north. Unfortunately, in addition to the latitude, we need another coordinate: the longitude.

Unfortunately, we absolutely need the longitude especially, if we want to go east to west on the ocean. This problem was so important that in the 18th century a substantial prize (US \$1.5M today) was to be awarded to the first person to produce a reliable method to find the longitude. In fig. 4.22, we see the Earth spinning along its North-South spin axis. The sphere is an extremely symmetric solid. The rotation provides an axis which uniquely identifies the equator. The latitude can be constructed by the set of all the circles parallel to the equator. An observer at point P on the Earth will possess the latitude given by the angle $\angle POR$, which we already stated can be determined by measuring the height of Polaris above the local horizon.

The other coordinate can be built by constructing the set of all the circles passing through the North and the South Poles. These circles, also-called *meridians*, all have the poles in common. There is not an easy method to determine the angle $\angle ROP$ because its zero is completely arbitrary and cannot easily be associated with sky objects. In a completely arbitrary way, the zero longitude is associated with the meridian passing through the city of Greenwich in England.

One way to determine the longitude would consist of determining the local midday, i.e. when the Sun reaches its highest altitude in the sky. If we had with us a clock that tells us the local time at Greenwich at the exact local midday, then we obtain the longitude by transforming the time difference in degrees. Since in one day there are 24 hours, then the longitude in degrees *at the equator* would simply be $L = \Delta t \cdot 360/24$ where Δt is the time difference

---

[15]Polaris is not *exactly* aligned with the Earth's rotation axis and is slightly less than a degree off ($0.66°$ in 2018).

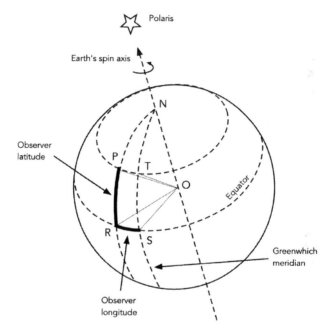

FIGURE 4.22 Spherical coordinates of an observer located at point P on the surface of the Earth. The latitude varies between −90° (at the South Pole) and +90° at the North Pole. The longitude is measured from 0° to 180° east or west of the Greenwich meridian defined at 0° longitude.

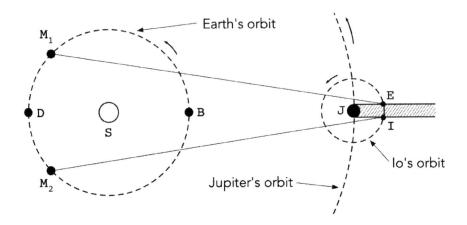

FIGURE 4.23 Geometry of the Sun-Earth-Jupiter system. The Sun illuminates Jupiter (J) creating a conical shadow behind. The innermost satellite Io enters and exits this shadow, respectively, at points I and E. When Io enters the conical shadow in I, its disappearance can be observed from the Earth along the curve DM_2B. The emergence from the shadow can be observed when the Earth is instead along the path BM_1D. When the Earth is in the special points B and D, Jupiter is said to be, respectively, in opposition and in conjunction.

in hours between the local position and Greenwich. If the result is bigger than $180°$ the longitude is $\ell = 360° - L$ in the west, otherwise the longitude is $\ell = L$ in the east direction. The last step would be to correct for the latitude by multiplying for the cosine of the latitude.

We have therefore seen that the problem of determining the longitude is equivalent to finding a good time-keeping device. Unfortunately, before the realization of accurate clocks, mariners did not have this technique. For this reason early astronomers proposed to use the apparent regular motion of planets and satellites to provide such an accurate clock. Galileo was one of the first to make such a suggestion when he discovered that Jupiter has satellites orbiting around it. The innermost satellite Io, in particular, has an orbiting period of only 42.5 hours, which must be extremely stable. After all, it has been orbiting Jupiter for the last few billion years.

The regularity of Io's orbits can be used as a "tick" of a cosmic clock. In particular, the appearance and disappearance of Io when it emerges or enters Jupiter's shadow, constitutes a very regular time-keeping device. Obviously this tick is very slow: instead of once per second it is once every 42.5 hours. For this reason a few astronomers in the past have spent many days observing Jupiter and its satellites taking a lot of observations of the orbits. Giovanni

Domenico Cassini, an Italian astronomer, was the first to use the timing of the eclipse of Jupiter's Moons to measure the longitude exactly in the way that Galileo suggested. Cassini managed to measure the longitude difference between Paris and the island of Hven near Copenhagen where he sent one of his colleagues. By measuring the difference in local times of the moment of the same eclipse of Jupiter's moons Cassini, correctly determined the difference in longitude between the two sites.

So everything was fine until the astronomical measurements started to be of good enough quality to see anomalies. In fact, a remarkable phenomenon appeared: the apparent ticking of the orbit of Io was not regular, but it changed in a strange way. When the Earth was receding from Jupiter, the timing of the orbit seemed to be slowing down, while, when the Earth was approaching Jupiter, the timing of the orbit seemed to be accelerating. The differences were remarkable and of the order of several minutes. This was a mystery because there was nothing more regular than the motion of planets or satellites. Think, for example, of the motion of the Moon around the Earth or the motion of the Earth around the Sun. These motions are extremely regular and predictable. There is nothing in Newton's equations that can explain the irregularities observed in the timing of the disappearance (or reappearance) of Io when entering (or exiting) its shadow with respect to the Sun (see fig. 4.23). How could the motion of the Earth around the Sun influence the orbit of Io?

The astronomer Ole Roemer gave an interesting explanation of the timing anomaly. In order to understand it, let us tell the story of Jim and Jane. Jim travels very often and in order to communicate with Jane he brings with him a cage full of carrier pigeons. Jim agreed that every day, always at the same time, he would send a message to Jane using one of the carrier pigeons. Jane does not know how far away Jim is and at what speed the pigeons fly. Jane knows that Jim will be moving along a straight line either towards her or away at a constant speed of 10 km/hour. Let's call T the time interval according to Jim of sending the pigeons: in this example it is $T = 24$ hours. Let's call c the speed of the pigeons and v Jim's speed with respect to Jane.

Day 1: Jane receives the first pigeon at 22:00. She does not know how far away Jim is because she does not know at what time Jim sent the pigeon. However, she knows that the day after, Jim will send another pigeon at exactly the same time. So the period T, i.e. the time interval between sending pigeons, is exactly 24 hours.

Day 2: Jane receives the letter at 17:36 and she immediately says: "I know that Jim is moving towards me and I also know at what speed the pigeons are flying!" How is this possible?

If Jim was not moving at all, Jane would have received the letter at exactly the same time as the day before, i.e. 10:00 pm. Jane does not know how far away Jim is.

In order to solve the problem, Jane does something clever: she assumes that Jim is exactly at the distance $d = cT$ where, as we said before, c is the speed of the pigeon and $T = 24$ hours is the time interval between sending

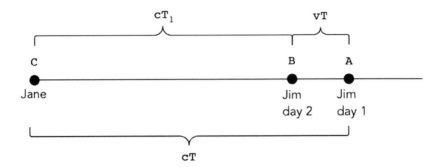

FIGURE 4.24 Jim sends a letter to Jane from location A on day 1. After exactly 24 hours Jim sends the second letter to Jane from position B on day 2. Being closer, Jane will receive the letter in less than 24 hours.

two letters by Jim. In fig. 4.24 we see the geometry: Jane is at point C and she is assuming that on day 1 Jim, at position A, has sent the first letter. The distance that the pigeon had to travel is therefore cT. On day 2, Jane receives the letter earlier than the day before, even though Jim has waited exactly 24 hours. Since Jim has traveled towards Jane, now Jim is at point B and the pigeon has to travel less distance cT_1. The distance between the point A and B is equal to vT. It follows that:

$$\overline{AC} = \overline{AB} + \overline{BC}$$
$$cT = vT + cT_1 \tag{4.36}$$

Jane knows that $T = 24$ hours, that Jim travels at $v = 10$ km/h, and she has recorded the time of arrival of the pigeon $T_1 < T$. She can now obtain the speed of the pigeon:

$$c = \frac{vT}{T - T_1} \tag{4.37}$$

We leave it to the reader to plug in the numbers and verify that the pigeon is flying at a speed of $c = 100$ km/h[16]. Notice that with this method, Jane cannot find the distance $\overline{AC}$, since she does not know at what time Jim is sending the pigeon.

Is it possible that Io's anomaly is due to the fact that the speed of light is finite? Can we infer the speed of light from the deviations observed in the timing of Io's occultations? What we just described is a well-known phenomenon in physics called the Doppler effect. In fact, the Doppler effect can be invoked to explain Io's anomaly [4].

---

[16] It seems that the average speed of carrier pigeons is around 97 km/h.

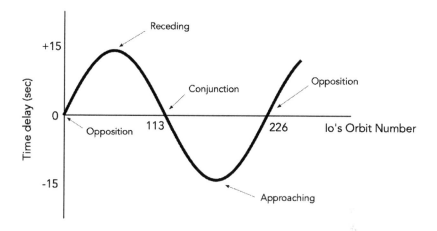

FIGURE 4.25 Deviation from the theoretical timing of Io's occultations with respect to the measured ones. When Io is receding from Earth, the timing appears to be slowing down. When it is approaching Earth, instead it seems to be accelerating.

Let's go back to fig. 4.23. If we record the deviations from the calculated real timing of the immersion or emersion of Io into/from Jupiter's shadow, we obtain a plot resembling fig. 4.25 adapted from [4]. If we start numbering Io's orbits when the Earth is in opposition to Jupiter (point B), the system Jupiter/Io has zero velocity with respect to Earth, i.e. the Earth is moving exactly at 90° with respect to the axis connecting the Sun S and Jupiter J. From now on, the Earth will be moving counterclockwise along the Earth's orbital path $\overline{BM_1D}$. Along this path, the Earth will be receding from the system Jupiter/Io and the immersion events of Io in Jupiter's shadow will appear to happen with increasing delay according to fig. 4.25. The delays in this figure are delays between successive rotations of Io. So if we keep track of the number of rotations, we can see that the delays will accumulate until orbit number 113. It turns out that after about 113 revolutions of Io around Jupiter, the Earth is now in conjunction at D and the accumulated delay is about 990 seconds. The distance $\overline{BD}$ is equal to twice the mean distance from the Earth to the Sun, i.e. twice the so-called Astronomical Unit = 149,597,871 km. We finally obtain, with this method, that the speed of light is about 300,000 km/s.

Think About It...

In the vicinity of black holes, tidal forces are so strong that objects are subject to extreme vertical stretching and horizontal pressure. Any object falling into a black hole will therefore be stretched into a long thin shape, thus the term spaghettification.

FURTHER READING

Cartwright, D.E. (1999), *Tides. A Scientific History.* Cambridge University Press.

Lewis, J.S. (1997), *Physics and Chemistry of the Solar System.* Academic Press.

Mazer, A. (2010), *The Ellipse. A Historical and Mathematical Journey.* Wiley.

Spence, J.C.H. (2020), *Lightspeed: The Ghostly Aether and the Race to Measure the Speed of Light.* Oxford University Press.

Bibliography

[1] *The Astronomical Almanac 2020.*

[2] E. J. Aiton. How Kepler discovered the elliptical orbit. *The Mathematical Gazette*, 59(410):250–260, 1975.

[3] Apollonius. *Conic Sections.* 200 BC.

[4] V. M. Babovic, D. M. Davidovic, and B. A. Anicin. The Doppler interpretation of Romer's method. *American Journal of Physics*, 59, 1991.

[5] A. E. L. Davis. Some plane geometry from a cone the focal distance of an ellipse at a glancet. *The Mathematical Gazette*, 91(521):235–245, 2007.

[6] Euclid. *Elements.* 300 BC.

[7] D. L. Goodstein and J.R. Goodstein. *Feynman's lost lecture.*

[8] Peter Lynch. The equation of time and the analemma. *Irish Math. Soc. Bulletin*, (69), 2012.

[9] I. Newton. *Philosophiae Naturalis Principia Mathematica.* 1687.

[10] A. M. Nobili and A. Anselmi. Relevance of the weak equivalence principle and experiments to test it: Lessons from the past and improvements expected in space. *Physics Letters A*, 382:2205–2218, 2017.

[11] Robert Weinstock. Newton's principia and the external gravitational field of a spherically symmetric mass distribution. *American Journal of Physics*, 52(10):883–890, 1984.

Index